Tobias Schwarze

Compact Environment Modelling from Unconstrained Camera Platforms

Schriftenreihe
Institut für Mess- und Regelungstechnik,
Karlsruher Institut für Technologie (KIT)

Band 040

Eine Übersicht aller bisher in dieser Schriftenreihe erschienenen Bände finden Sie am Ende des Buchs.

Compact Environment Modelling from Unconstrained Camera Platforms

by
Tobias Schwarze

Scientific
Publishing

Dissertation, Karlsruher Institut für Technologie
KIT-Fakultät für Maschinenbau

Tag der mündlichen Prüfung: 04. Oktober 2017
Referenten: Prof. Dr.-Ing. Christoph Stiller
Prof. Dr.-Ing. Fernando Puente León

Impressum

KIT Scientific Publishing

Karlsruher Institut für Technologie (KIT)
KIT Scientific Publishing
Straße am Forum 2
D-76131 Karlsruhe

KIT Scientific Publishing is a registered trademark
of Karlsruhe Institute of Technology.
Reprint using the book cover is not allowed.

www.ksp.kit.edu

Print on Demand 2018 – Gedruckt auf FSC-zertifiziertem Papier

ISSN 1613-4214
ISBN 978-3-7315-0801-4
DOI 10.5445/KSP/1000083235

Compact Environment Modelling from Unconstrained Camera Platforms

Zur Erlangung des akademischen Grades

Doktor der Ingenieurwissenschaften

der Fakultät für Maschinenbau
des Karlsruher Instituts für Technologie (KIT)
genehmigte

Dissertation

von

DIPL.-ING. TOBIAS SCHWARZE

aus Hildesheim

Tag der mündlichen Prüfung: 4. Oktober 2017

Hauptreferent: Prof. Dr.-Ing. Christoph Stiller
Korreferent: Prof. Dr.-Ing. Fernando Puente León

Acknowledgement

The work at hand was written during my time at the Institute of Measurement and Control Systems (MRT) at the Karlsruhe Institute of Technology (KIT). It was supervised by Prof. Dr.-Ing. Christoph Stiller whom I thank for the opportunity to complete this research and for providing the stimulating environment with various excursions, seminars and conference visits. I would also like to thank Prof. Dr.-Ing. Fernando Puente León for agreeing to be the second reviewer of this thesis.

Many people have made this project a memorable time. I would like to thank Dr. Martin Lauer for many fruitful discussions and suggestions, for proofreading articles and parts of this work and for the opportunity to get involved in the exceptional world of visually impaired people. For the close cooperation within the OIWOB project I thank Manuel Schwaab, Michailas Romanovas, Sandra Böhm and Prof. Dr.-Ing. habil. Thomas Jürgensohn. Furthermore, I thank all my colleagues, who contributed with numerous discussions in coffee breaks and seminars and joined diverse spare time activities. Especially I thank Johannes Gräter, Eike Rehder and Hannes Harms for proofreading parts of this thesis and their valuable feedback.

Many students have invested time and influenced this work, I would like to thank them again at this point. Thanks to our mechanical workshop with Günter Barth and Goran Cicak, who did not get tired of mounting cameras into various uncommon platforms, to Werner Paal for providing excellent IT support, and to Sieglinde Klimesch, Erna Nagler and Ines Rapp for hiding bureaucracy. I am grateful for the financial support received from the Karlsruhe School of Optics and Photonics (KSOP).

Last I wish to thank my family and my companion Karien, whose steady support I could always count on!

Karlsruhe, April 2017 *Tobias Schwarze*

Abstract

Mobile robotic systems need to perceive their surroundings in order to act independently. They need to determine the space of safe movement and detect obstacles and understand their motion. To this end they are equipped with sensors which provide information about the unknown environment. By interpreting sensor measurements, a representation of their environment arises that provides the relevant information in an accessible way. Mobile systems are subject to constraints that render this perception process challenging and unsolved in many aspects. Hardware and sensors must be small, lightweight and energy efficient while providing perception ranges as wide as possible. The requirement of minimal processing times conflicts with clear computational performance limits. In this work we present a perception framework that meets these requirements and builds upon the versatile possibilities of a binocular camera as sensory input.

The perception framework is neither limited to a specific task, nor to a specific platform. Only minimal assumptions are made regarding the potential motion or orientation of the cameras. This allows for the application to unevenly walking robotic systems or also for completely passive sensing e.g. with the cameras attached to a human wearer. The framework transforms the raw camera data into a compact meta representation consisting of instances of arbitrary objects. By strongly abstracting from details, a compact model of the environment is formed, which emphasizes the essential information. During the modelling only little assumptions can be made regarding the objects that surround the system. Classical methods to detect objects of specific categories are infeasible and are therefore replaced by means of different scene cues. We introduce a number of different algorithms which complement each other to first explain the static scene background structure and subsequently model generic objects and their motion in the scene foreground.

For autonomous mobile systems the representation is an efficient basis for subsequent algorithms that e.g. perform path planning. The abstract scene model is immediately applicable for collision avoidance and targeted navigation towards or around objects, and enables to orient along background structures like building facades. The applications are not limited to closed technical systems. The compactness of the representation allows to efficiently communicate the surrounding situation to a human user. In an experiment we show applications of augmented reality, for instance in entertainment or education. Based on the environment modelling we develop a new kind of technical assistance system for visually impaired persons. The high level of abstraction enabled an acoustic feedback design, which informs the user about hazards in the surrounding and is intuitive to use without lengthy training periods. An experimental study shows how visually impaired users can benefit from such system.

Kurzfassung

Mobile Robotersysteme müssen ihre Umgebung wahrnehmen, um selbstständig agieren zu können. Sie müssen den freien Raum bestimmen, in dem sie sich sicher bewegen können, und Hindernisse und deren Bewegung erkennen. Sensoren ermöglichen ihnen dabei die Erfassung ihrer Umwelt. Durch Interpretation der Sensordaten entsteht ein Modell der Umgebung, das die relevanten Informationen auf geeignete Weise bereitstellt und zugänglich macht. Mobile Systeme unterliegen einigen Einschränkungen, welche diesen Prozess der Wahrnehmung erschweren und in vielerlei Hinsicht zu einem ungelösten Problem machen. Ihre Hardware und Sensorik muss klein, leicht und energieeffizient sein und dabei einen möglichst großen Wahrnehmungsbereich abdecken. Die eingeschränkten Rechenleistungen mobiler Hardware stehen im Konflikt mit erforderlichen minimalen Verarbeitungszeiten. In dieser Arbeit stellen wir ein Wahrnehmungsframework vor, das diese Voraussetzungen berücksichtigt und auf den vielseitigen Möglichkeiten eines binokularen Kamerasystems aufbaut.

Das Framework ist weder auf eine bestimmte Anwendung noch auf eine bestimmte Plattform beschränkt. Es werden nur minimale Annahmen bezüglich der potentiellen Bewegung oder Orientierung der Kameras gemacht. Dies ermöglicht den Einsatz z.B. auf stark bewegten Laufrobotern oder auch eine komplett passive Umwelterfassung mit am Körper getragenen Kameras. Das Framework transformiert Kameradaten in eine Umgebungsrepräsentation auf Basis von Objektinstanzen beliebiger Art. Durch starke Abstraktion von Details entsteht ein kompaktes Modell der Umgebung, das die essentiellen Informationen herausstellt. Während der Modellierung können nur wenige Annahmen über die Objekte in der Umgebung getroffen werden. Klassische Methoden zur Detektion von Objekten bestimmter Kategorien sind nicht einsetzbar und werden mithilfe des Szenenkontexts ersetzt. Wir stellen verschiedene Algorithmen vor, die sich gegenseitig ergänzen, um zunächst den

statischen Szenenhintergrund zu ermitteln und darauf aufbauend beliebige Objekte und deren Bewegung im Szenenvordergrund modellieren.
Für autonome Systeme ist die Repräsentation eine effiziente Basis für nachfolgende Algorithmen z.B. zur Pfadplanung. Das abstrakte Szenenmodell kann unmittelbar zur Kollisionsvermeidung oder zur zielgerichteten Navigation eingesetzt werden und ermöglicht dabei die Orientierung entlang von Hintergrundstrukturen wie etwa Gebäudefassaden. Die Anwendungen sind nicht auf technische Systeme begrenzt. Die Kompaktheit des Modells ermöglicht es, die Umgebungsinformationen effizient an einen menschlichen Nutzer zu übermitteln. In einem Experiment zeigen wir Anwendungen der erweiterten Realität, zum Beispiel zu Unterhaltungs- oder Schulungszwecken. Auf Grundlage der Umgebungsmodellierung stellen wir ein neuartiges Assistenzsystem für Blinde vor. Das hohe Abstraktionslevel ermöglicht den Einsatz eines akustischen Feedbackdesigns, das den Nutzer ohne langwierige Lernphase intuitiv verständlich über Gefahren in der Umgebung informiert. Eine experimentelle Studie zeigt abschließend den potentiellen Nutzen eines solchen Systems.

Contents

Notation and Symbols

Acronyms

DATMO	**D**etection **A**nd **T**racking of **M**oving **O**bjects
GPU	**G**raphics **P**rocessing **U**nit
HMD	**H**ead-**M**ounted **D**isplay
HRTF	**H**ead-**R**elated **T**ransfer **F**unction
IMU	**I**nertial **M**easurement **U**nit
MEMS	**M**icro**e**lectro**m**echanical **S**ystem
RANSAC	**Ran**dom **Sa**mple **C**onsensus

Symbols and General Notation

$\mathbf{P}^L$	Euclidean point in coordinate frame L
T_W	Pose in coordinate frame W
(u, v)	Pixel position in image space
$\delta(u, v)$	Disparity at pixel position (u, v)
$uv\delta$ point	Point in image space with corresponding disparity
$F(\cdot)$	Inverse camera projection function
B	Camera baseline
f	Camera focal length
(c_u, c_v)	Camera principal point
K	Intrinsic camera matrix
M	Extrinsic calibration between camera and IMU
$\mathbf{x}_k^-$	A priori estimate of state $\mathbf{x}$ at time k
$\mathbf{x}_k^+$	A posteriori estimate of state $\mathbf{x}$ at time k
$\Sigma_{\mathbf{P}}^W$	Covariance of $\mathbf{P}$ in coordinate frame W

1 Introduction

Mobile robotic systems are expected to become a natural part of our everyday environments in midterm future. Autonomous cars are just one prime example for such systems. Advances in computational engineering, battery and sensor technology provide a promising basis for the research done within the last decades. Already today it has matured into intelligent autonomous systems that transport goods, guide tourists, mow the lawn, inspect bridges, or alert of intruders.

In order to operate in unknown environments, such systems need to perceive their surrounding. For safe and targeted navigation they need to be aware of obstacles, dangers and potential objects of interest. It allows them to derive the space of safe movement, derive a navigation strategy, or also to interact with their environment. Environments are very diverse and complex, especially in urban settings. Besides the immense variety of different scene elements, these environments are subject to frequent change. Dynamic objects are moving through the scene and temporally occlude parts of the environment and other objects. Additionally, the systems themselves are moving. As a result, not only their environment, but also their sensory impression of it is constantly changing. In such environments, the systems need to be aware of the presence of objects and their potential movements, but also of their own movement. All information has to be derived from the raw data of sensors, that connect the system with their surrounding world. The interpretation of sensory information into a comprehensible and useful representation is the challenging problem of perception.

Mobile systems exhibit a number of limitations. Their small design and construction constrains the size and weight of computational hardware, batteries and sensors. This leads to limited computational capacities, limited operation time, and sensors with low bandwidth and accuracy. Perception methods need to be robust, while computationally lightweight in order to process data in real-time on hardware that is pared down for energy efficiency. This thesis presents a framework to perceive the environment of a mobile system under these conditions using solely the input of a binocular camera.

To facilitate high-level tasks like behavioural planning, most mobile robotic systems apply multiple stages of abstraction to the low level sensor data. They create a meta representation that contains the sensed information in a semantically enriched and better accessible form. Besides representing the current sensor readings the representation may incorporate past observations, so that it forms a model of the surrounding environment. The modelling is usually geared to specific tasks or situations and differs accordingly in expressiveness, compactness and its semantic level. Our physical world is made up of objects that we have learned to group into concepts and categories, which share certain attributes and enable high-level thinking. For artificial reasoning processes a similar degree of semantic information would be desired. Yet, unless explicitly modelled or learned, this abstract concept knowledge is not available to a machine. For many tasks such high level of information is not required, but a semantic level of general object instances is sufficient. Consider a collision avoidance scenario, for which it is irrelevant at first, whether an approaching obstacle is a static pole, or a dynamic cyclist. However, it is advantageous to recognize a pole or a cyclist as individual instances of some arbitrary concept. Such a level of object instance modelling provides a basis for reasoning processes. It leads to extremely compact environment representations, which abstract from the details and highlight the essential information. However, this modelling process is very difficult. Since nothing is known about the objects, hardly any assumptions can be made to facilitate their recognition and separation from the scene. Especially the typical object shape or size, typical motion or typical appearance would be valuable prior knowledge. Moreover, without referring to known concepts, it is not even clear what exactly constitutes a valid object.

The methods proposed in this thesis aim at modelling the environment on such level from the input of a binocular camera. A meta representation is found that is useful in different application scenarios. The modelling concerns a local area around the system, in which individual object and obstacle instances are detected and modelled in temporal as well as spatial manner. These enable a system to carry out tasks like close range navigation, collision avoidance with moving objects, up to interaction with objects. The system itself must not be static and might be subject to passive motion that is beyond its control.

1.1 Environment Representations

In mobile robotic systems, meta representations are found in different forms. They can represent data of a single sensor reading, but they can also represent data of possibly multiple different sensors accumulated over time and different sensor poses. These latter representations often resemble a spatial map of the environment. The system performs a mapping of sensor data into the environment model. To this end it estimates its pose with respect to the model. Thereby the model can be extended over time, and past information can be accessed, which is not in sensor range any more. This facilitates to model scene structures that are larger than the actual sensor range. Figure 1.1 shows two examples of environment models, that are often found in the context of traversability analysis.

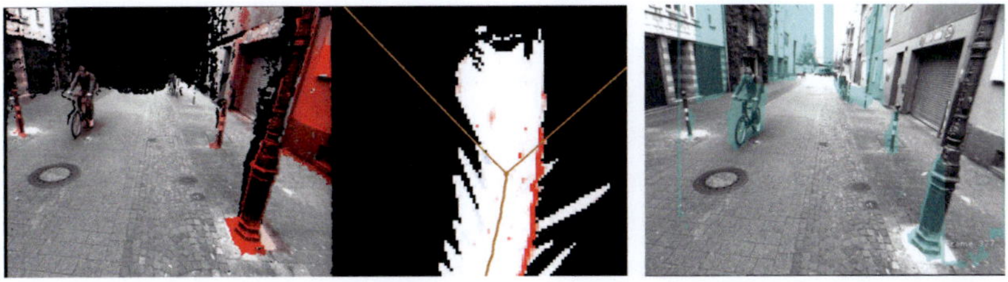

Figure 1.1: Left: Scene representation in form of a two-dimensional floor occupancy map (top view). Red color denotes elevated parts of the scene, white cells are free. Cells are updated over time with the current measurements and the estimated sensor movement. **Right:** The intermediate *Stixel* representation models the limit of free space using the current sensor data.[1]

[1] The implementation was kindly provided by Martin Lauer (occupancy map) and Hannes Harms (Stixel)

An essential design criterion is the requirement of topological model consistency. A mobile system that revisits a previously visited place does not know about the topological connection to the previous visit. Only if it recognizes the place this connection can be established. For systems that perform active exploration a topologically consistent model is of utter importance. For other tasks, which are primarily concerned about the local environment, it is dispensable. The ability of recognizing these events termed loop-closure comes at the cost of storing and maintaining a model of the entire visited environment. Though the density of the actual representation can greatly vary from sparse landmarks to dense 3D point clouds or closed surfaces, the problem of scalability is not easy to overcome. Where global consistency is not important, a locally consistent model of the environment is much less resource consuming. Past data can simply be forgotten without affecting the functionalities that the model fulfills locally around the system.
An assumption made during the basic mapping methodology is that the environment is static. Dynamic parts of the environment are treated as model violations and need to be explicitly detected in order to avoid dynamic objects from being mapped spuriously. However, when moving in populated environments, dynamic objects play a key-role. Latest when planning a collision free route, the dynamics of moving objects need to be modelled in addition to the static scene. Otherwise, route planning may completely fail if solely the current freespace is considered [23]. The importance of dynamic objects has led to several extensions that target the *detection and tracking of moving objects* (DATMO) in conjunction with mapping the environment. The result is a map of the static environment, that is augmented with independently moving objects [91]. It allows to move freely while avoiding collision with the static scene and other moving objects, or also allows to follow these objects, as long as they are moving.
As soon as interaction with the environment is required, the environment must be understood on a higher level. Beyond sensing free space, it becomes important to explain what is limiting the free space. Such level of scene understanding provides very valuable context knowledge to higher level tasks. This potential is well recognized and has led to various approaches to augment unstructured maps with semantic information.

Figure 1.2: Examples of environment models enriched with semantic information: **Left:** Floor plan with room type labeling [96]. **Top:** Semantic road scene reconstruction [90]. **Bottom:** Instance mapping of known objects [76].

Examples include labeling indoor maps with room types [96], or reconstructed road scene models with semantic concepts, e.g. [90, 82] (see Figure 1.2). Maps that are semantically labeled in such manner provide spatial context information. They state whether the system is positioned on a pavement or on the road, whether an area ahead is a large water basin or drivable lawn, or state that the current room is a kitchen. Such context strongly supports object recognition [66]. Many higher level tasks like object interaction require information on a level of concept instances. A semantic map will contain the information that certain parts of the map are cars, but not how many and what the different instances of cars are. This kind of information requires a decomposition of the map into concept instances – or objects.
When object instance perception is required, the most common solution is to teach the system about the typical appearance [14, 70], or provide a geometric model [76] of the required concept (Figure 1.2). This facilitates direct instance recognition, e.g. by sliding window classification or model fitting. Unfortunately, the principle does not generalize very well and is infeasible in urban environments with a large number of different object concepts. The earlier introduced DATMO methodology works around this problem by detecting objects by their movement. In general though, objects should be modelled as such whether they are moving or not. This is a keynote of the perception framework that is proposed in this work.

1.2 Contribution

The proposed methods abstract raw sensor data of a binocular camera and create a high-level spatio-temporal object instance representation, which is useful in a variety of applications (Figure 1.3). Besides its potential for high level interaction tasks and reasoning, an object instance level perception allows efficient representations of the environment through geometric primitives. A wall, once recognized as a wall, can be expressed as a geometric surface with few parameters, compared to thousands of low level measurements in an unstructured model as e.g. in a metric point cloud. The large amount of information that surrounds the system becomes strongly condensed. This results in an extremely compact representation well suited for mobile platforms where limited computational resources have to be handled economically. On algorithmic side, maintaining an environment model of geometric primitives alters from a plain mapping to complex registration and multi-model estimation problems.

To this end, several well-matched algorithms are introduced: The concept of vanishing points is exploited to gather information about the scene geometry. An efficient algorithm is developed that maintains an accurate model of current prominent scene directions while the system traverses through the scene. This knowledge is utilized in different ways. First, a method is presented to track geometric surfaces under the special conditions of full occlusion and invisibility. Of particular interest is the navigable ground surface, which is widely assumed to remain in perception range permanently. With free moving, uncontrolled camera setups this assumption is frequently violated. Vanishing points are shown to provide a strong geometric feature to handle these situations. Afterwards, an algorithm is introduced that extends the range of operation to multi-floor environments by specifically modelling stair transitions. Here, vanishing points provide the stair and step orientation during traversal.

The focus of the proposed representation are tasks within the local environment. Other than the related field of environment mapping, global model consistency and topological correctness are ignored consciously. However, a method is presented that makes use of the modelled scene information in order to extend the consistency to the surrounding scene.

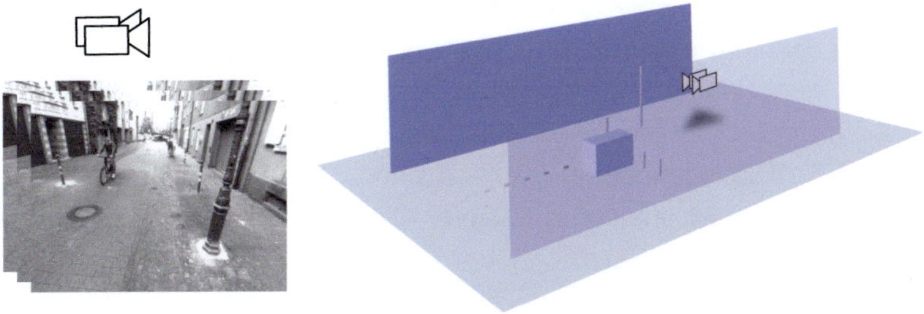

Figure 1.3: The system transforms raw data from a binocular camera into an abstract environment model on a semantic level of object instances.

Finally, a method is presented which detects objects of arbitrary kind and models them spatially as well as temporally. A tracking formulation is presented that simultaneously estimates shape and motion of surrounding objects. This enables to track objects through occlusions and after they have left the visible sensor range.

Strongly limited computational capabilities and physical constraints of mobile systems require several trade-offs, which all algorithms have to cope with. Sensors for mobile systems need to be small and lightweight while covering a sensing range as large as possible. Short binocular camera baselines and wide angle lenses with short focal length contribute to large uncertainties in depth measurements. Low image resolutions are inevitable in order to keep processing times within tight limits. All algorithms are trimmed to be applicable in real-time on standard hardware without demand of specific hardware like graphic processors (GPUs). To exploit the potential of current multi-core platforms, a parallelized software architecture is developed, which ensures small latencies while keeping processing rates as high as possible.

Little assumptions are made regarding the camera motion. This allows to deploy the methods also in unconventional settings, as for instance unevenly walking robotic platforms, or cameras attached to a human wearer. This makes the framework interesting for virtual reality entertainment applications, or in the assistance of visually impaired persons. Exemplary applications are implemented and their potential use discussed within the experiments.

Organization In Chapter 2 we first recapitulate the basics of depth perception and egomotion estimation using a binocular camera, with a focus on relevant sources of errors and general weaknesses. This provides the basis for the proposed framework to model the environment which we detail in Chapter 3. The experimental platforms and considerations regarding the implementation are subject of Chapter 4. Two experiments demonstrate the potential use for wearable systems. Concluding remarks are given in Chapter 5.

2 Depth and Egomotion Estimation

The basis of the proposed framework are algorithms to estimate scene depth and the platform's motion from the binocular camera images. Efficient solutions exist for both problems. Estimating the motion that the cameras undergo allows to incrementally estimate the system's pose with respect to an initial starting point. The measured scene depth can be transformed to metric measurements. The combination of both algorithms enables to reconstruct scenes that are larger than the sensor range. A prerequisite is a camera setup with known intrinsic and extrinsic calibration. To facilitate the point correspondence search between both views, a rectification step virtually aligns the undistorted camera images horizontally on a common image plane. A general configuration of such system is depicted in Figure 2.1. A scene point in the Euclidean camera space L, denoted as $\mathbf{P}^L$, generates a measurement in image space $\mathbf{p} = (u, v, \delta)$ with pixel position (u, v) and disparity δ (termed $uv\delta$ point in the remainder). $\mathbf{P}$ is transformed to the scene fixed coordinate space W as $\mathbf{P}^W = T_W \mathbf{P}^L$. The transformation T_W is incrementally estimated by odometry measurements.

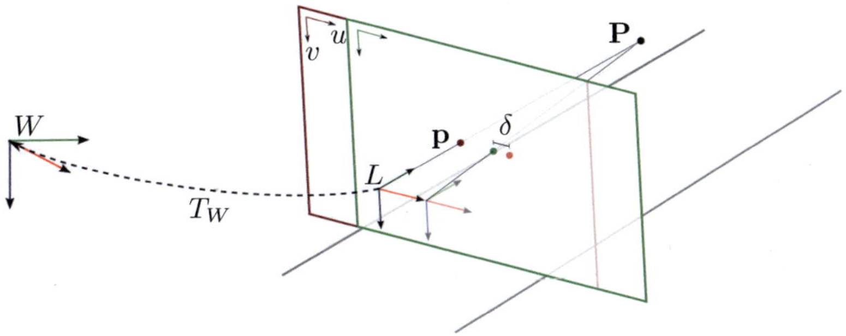

Figure 2.1: Relations between image space with measured point $\mathbf{p}$, Euclidean camera space L and scene fixed coordinate space W.

In problems of projective geometry the Euclidean space is usually represented in the system of homogeneous coordinates. A point in Cartesian coordinates $P = (X, Y, Z)$ is expressed in homogeneous coordinates with an additional coordinate h and becomes $\mathbf{P} = (hX, hY, hZ, h)$. The Cartesian coordinates are recovered as $(X/h, Y/h, Z/h)$. Consequently, the coordinates $\mathbf{P}$ and $k\mathbf{P}$ represent the same point for all non-zero values of k. This notation allows to represent points in infinity with finite values ($h = 0$). Points $(X, Y, Z, 0)$ lie on an ideal plane in infinite distance and are handled just like finite scene points. Besides this, homogeneous coordinates simplify the notation of transformations between coordinate systems. An affine transformation T with linear part R and translation $\mathbf{t}$ is written as augmented matrix $T = \left(\begin{smallmatrix} R & \mathbf{t} \\ \mathbf{0}^\mathrm{T} & 1 \end{smallmatrix}\right)$. The transformation of a point $\mathbf{P}$ is then achieved by the matrix multiplication $T\mathbf{P}$.

2.1 Depth Estimation and Reconstruction

In a pair of rectified binocular images, depth is measured as the image column offset between the projections of a scene point into the left and right image. The offset is commonly termed *disparity* and relates an image point (u, v) in the left image with the image point $(u + \delta(u, v), v)$ in the right image. Efficient algorithms exist to estimate the disparity for each image point of the first camera by seeking the corresponding image point in the image of the second camera.

The Euclidean camera coordinates of a $uv\delta$ point $\mathbf{p} = (u, v, \delta(u, v))$ are reconstructed via

$$\mathbf{P}^L = \begin{pmatrix} X \\ Y \\ Z \end{pmatrix}^L = F(\mathbf{p}) = \frac{B}{\delta} \begin{pmatrix} u - c_u \\ v - c_v \\ f \end{pmatrix} \tag{2.1}$$

where B is the camera baseline, f the focal length, and (c_u, c_v) the principal point that model a pinhole camera. The non-linear mapping causes the measurable depth resolution to decrease with increasing distance from the camera. Assuming a typical mobile setup with a baseline of 20 cm and an image column resolution of 640 pixels, 1 pixel disparity difference covers a range of more than 30 m in a camera distance of 30 m where $\delta \approx 1$. This range grows to ∞ for $\delta \rightarrow 0$. Typically, 8 sub-pixel disparities are estimated by non-linear refinement, which mitigates the effect to a distance resolution still as big as 1 m in 25 m camera distance. An approximation for the covariance S of reconstructed points is given by a Taylor approximation as

$$\Sigma = J_F \cdot M_{uv\delta} \cdot J_F^T \tag{2.2}$$

$$J_F(u,v,\delta) = \begin{pmatrix} \frac{dF_X}{du} & \frac{dF_X}{dv} & \frac{dF_X}{d\delta} \\ \frac{dF_Y}{du} & \frac{dF_Y}{dv} & \frac{dF_Y}{d\delta} \\ \frac{dF_Z}{du} & \frac{dF_Z}{dv} & \frac{dF_Z}{d\delta} \end{pmatrix} = \begin{pmatrix} \frac{B}{\delta} & 0 & \frac{-B(u-c_u)}{\delta^2} \\ 0 & \frac{B}{\delta} & \frac{-B(v-c_v)}{\delta^2} \\ 0 & 0 & \frac{-Bf}{\delta^2} \end{pmatrix} \tag{2.3}$$

where J_F is the Jacobian of $F(\cdot)$, and $M_{uv\delta}$ is the diagonal matrix of measurement variances $M_{uv\delta} = \text{diag}(\sigma_u^2, \sigma_v^2, \sigma_\delta^2)$.
In practice a trade-off between quality and measurement frequency has to be found. Figure 2.2 shows a visual comparison between two extreme algorithm configurations. Some errors appear independent of matching strategy, refinement or image resolution. These are of principal nature and concern image areas which lack texture information (e.g. caused by overexposure), depict very repetitive structures, or depict reflective surfaces, as e.g. on cars. All cases can cause enormous errors which are almost impossible to discern. Accordingly, algorithms that process disparity depth data have to be robust to such errors.

Left input image (a) OpenCV SGBM 640x480 pixel (b) libToast2 320x240 pixel no subpixel refinement

Figure 2.2: A visual comparison of disparity estimators where color encodes depth. Configuration (a) achieves a measurement frequency of 10 Hz (modified semi-global matching implementation [39]), configuration (b) lowers quality but achieves almost 100 Hz (block matching implementation [73])

2.2 Egomotion Estimation

Mobile systems that explore unknown environments usually do not have access to a global positioning reference. The platform position cannot be directly measured in such environments, but has to be estimated relative to some starting location by integrating measurements of the travelled distance. The estimated position is subject to the accumulating measurement error, also termed drift. Borrowed from the term odometry, the process of estimating the motion of a camera solely from the captured image data is commonly referred to as *visual odometry*.

The general goal of visual odometry is to find a rotation R and a translation $\mathbf{t}$ that relates the camera pose at time k to the camera pose at time $k-1$ through the rigid transformation

$$T_{k-1,k}(=T_k) = \begin{pmatrix} R & \mathbf{t} \\ \mathbf{0} & 1 \end{pmatrix}.$$

Most available methods are based on salient image features $\mathbf{x}$ which can be unambiguously matched across subsequent images. A geometric distance error between the feature point locations $(\mathbf{x}_{k-1} \leftrightarrow \mathbf{x}_k)$ at times $k-1$ and k is minimized to find T_k. This distance error is typically defined as a reprojection error in image space, a metric error in Euclidean space, or a mixture of both. In the monocular camera case, the translation part $\mathbf{t}$ remains principally unscaled. A recent overview about the sub-problems can be found in [77].

A byproduct of feature based visual odometry is the motion compensated scene flow between feature matches $(\mathbf{x}_{k-1} \leftrightarrow \mathbf{x}_k)$. It results from forward projecting the image feature points $\mathbf{x}_{k-1}$ with the estimated transformation T_k. The compensated feature match becomes $\left(F^{-1}\left(T_k \cdot F(\mathbf{x}_{k-1})\right) \leftrightarrow \mathbf{x}_k\right)$. It vanishes for static scene parts, and represents the motion of moving scene parts as if measured from a static camera at time k.

Similar to disparity estimation, feature based visual odometry depends on sufficiently textured scenes to generate point correspondences across images. The scene has to contain apparent static parts and camera motion has to be small enough to ensure overlap with the previously captured image.

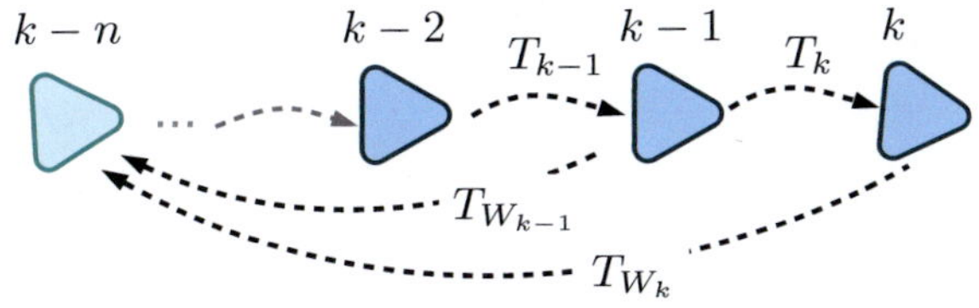

Figure 2.3: The global transformation T_W is incrementally estimated through odometric measurements T_k.

This can be met by high measurement rates, which come along with a high computational load.
Usually one is not only interested in the camera pose to pose motion, but also in the overall travelled path, or the current pose T_{W_k} at time k with respect to the initial global coordinate frame at time $k-n$. This transformation incrementally accumulates as visualized in Figure 2.3 to

$$T_{W_k} = T_{k-n+1}^{-1} \cdots T_{k-1}^{-1} \cdot T_k^{-1} \tag{2.4}$$

$$= T_{W_{k-1}} \cdot T_k^{-1}. \tag{2.5}$$

Because of its incremental nature this estimation is inherently subject to drift, caused by errors made in the estimation of T_k which accumulate over time (Figure 2.4). Different strategies exist to minimize this drift. When features are tracked over multiple frames, drift can be reduced by windowed bundle adjustment which jointly optimizes feature locations and camera poses over the last few estimation steps [89]. However, drift can still grow unlimitedly. To achieve a global bound, the earlier mentioned concept of loop-closure detection has to be incorporated. A different strategy consists of measuring a global property of the scene. For instance, inertial aided odometry uses the direction of gravity and the magnetic north to gather referenced tilt and heading measurements. These can be correlated with the odometry to correct the small estimation errors in the rotational part. This approach will be detailed in Section 3.1.5, where it is applied with solely visual references.

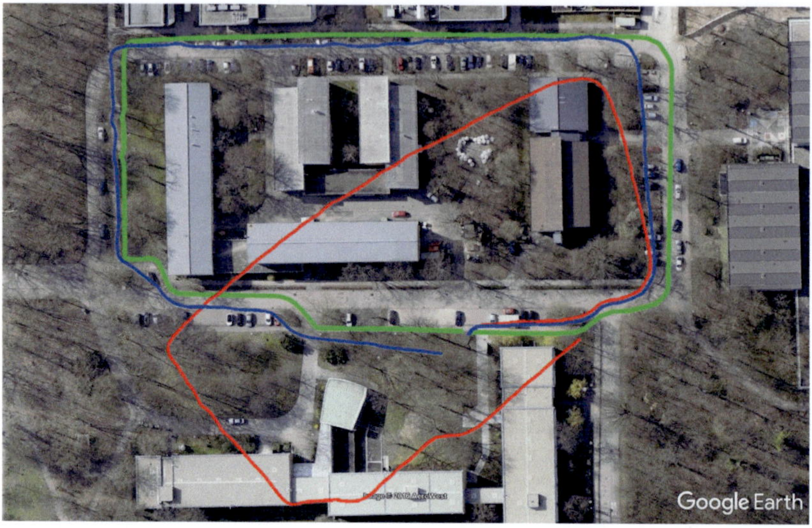

Figure 2.4: Odometric position estimation is afflicted with drift. The red path is a visual odometry estimate of the true path indicated in green. The sequence is captured with a head-worn binocular camera during walking. The cameras are subject to strong egomotion which leads to increased inaccuracies when compared to e.g. wheeled platforms. A visual inertial filter can compensate for the drift by incorporating the referenced heading and tilt measurements of an inertial measurement unit (blue path) [100]. Background image: Google Earth, © 2016 AeroWest.

3 Visual Scene Perception

While a mobile system is moving through an environment it is faced with large amounts of possibly different kinds of sensor data. The task of the perception framework is to fuse these measurements and convert them into a representation which is accessible and useful for the envisioned application. This representation, referred to as environment model, facilitates the tasks of the system. A very common such task could be movement in unknown environments.

Safe navigation involves two aspects: On one side, a high-level routing problem must be solved that leads to a certain goal. In unknown environments and limited perception range (consider 20-30 meter with a mobile camera setup), this compulsorily reduces to a lower-level local path planning problem that leads into a certain direction. Man-made environments provide many cues that are helpful at this. Consider for example the direction of a curbstone that separates street from sidewalk, or the alignment of building facades. Afterwards, the actual movement into the given direction has to be performed. Different objects might be placed on the path, large objects might extend into the path, persons might be moving across or along. Avoiding collision with such obstacles, or heading towards objects of interest is the other side of safe navigation.

The environment model must provide information that enable to handle these situations. Certainly, path planning in static environments is a prime example for environment representations that model the free space. However, reasons of expressiveness, meaningfulness and compactness advocate a representation that instead focuses on the objects that constrain the navigable space. It is a step towards higher-level scene understanding and offers interactive opportunities far beneath path planning. This chapter deals with the questions, how such representation could look, and how it can be derived from the data provided by a binocular camera system.

Without embedding prior knowledge about the world, typical top-down reasoning from concepts to objects (i.e. based on features like appearance, shape, size, etc.) remains precluded. Other features are required that allow a bottom-up modelling process from sensor data to object instances. General conditions to form an object could be summarized as follows:

(a) Objects are not traversable and need to be avoided.

(b) Objects feature a limited size and a clear boundary, so they can be circumnavigated.

(c) Objects might be moving, but they might as well be static.

(d) Objects are meaningful in a way that they represent a concept on a medium level of abstraction.

(e) Objects are neither a mixture of different such concepts, nor merely a part of such.

A wheel on a car is part of the concept *car*, and not an own object instance. In contrast, the same wheel lying on the road is by itself a valid object instance. This toy-example points up the difficulty in the mere definition of an object. It is highly context dependent. To dissolve ambiguities high-level knowledge is required, but not available.

Looking at urban environments it is striking that large parts of the scene cannot be explained using these conditions. Man-made environments feature large buildings, fences and hedges, stairways, etc. All are either of unclear extent or theoretically well traversable, and always immovable. Rather than being objects, these structures can be understood as a limiting frame of the scene, consisting of an arrangement of surfaces. As Gibson noted in his psychological view on perception, "the impression of a continuous surface may account for visual space conceived as a background" [31], it provides a background context, in front of which objects form the scene foreground. Background structure plays an important role during scene perception.Gibson further noticed that "there is literally no such thing as a perception of space without the perception of a continuous background surface".

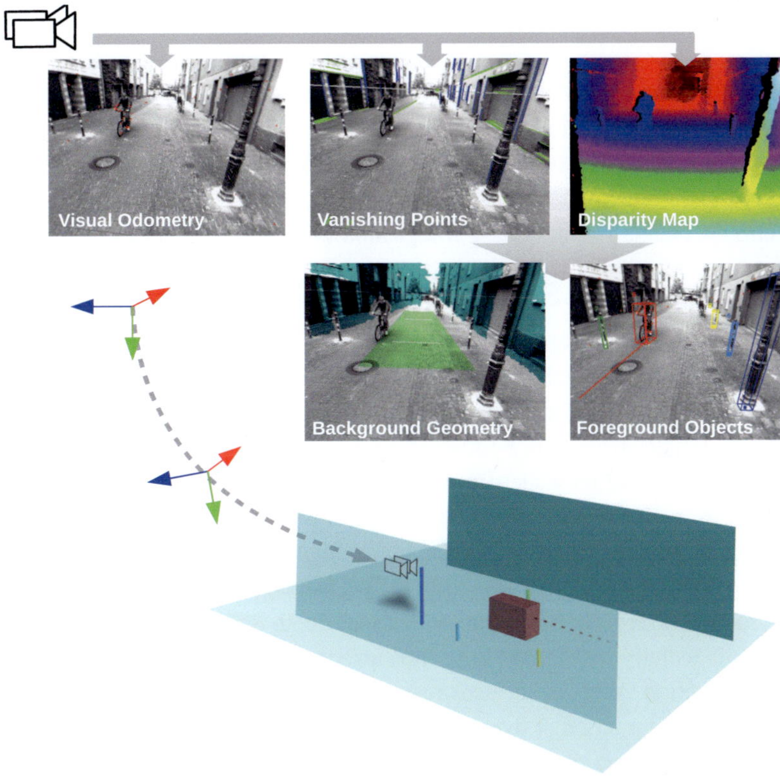

Figure 3.1: High-level overview of the applied algorithms.

Background surfaces allow the impression of distance without measuring depth, and provide important orientation clues. Of special importance in this context is the ground – a mobile system without ability to perceive ground orientation can be considered completely disoriented.
The task of the perception framework is to build and maintain the environment model consisting of background structure and foreground objects while moving through the scene. The sensor range will typically only cover a fraction of the surrounding environment. New measurements need to extend or update existing parts of the model. An overview is provided in Figure 3.1: All

measurements are derived from the binocular camera system. The disparity depth map provides metric measurements of the scene, while the egomotion estimation in terms of visual odometry ensures a locally consistent modelling in spatial and temporal regards. As an additional mid-level feature to measure prominent scene orientations vanishing points will be introduced. These three features provide the basis for modelling the scene background structure (Section 3.1), and for modelling foreground objects (Section 3.2).

3.1 Geometric Scene Background

The most obvious background feature is the ground floor, which constitutes the theoretically navigable space. Further delimiting structures are found with buildings, walls, fences or hedges, which are usually vertically aligned with the ground surface. In a limited area around the point of view, all these can be seen as continuous surfaces and as such be modelled geometrically. Geometric surface models are attractive in terms of compactness and because they efficiently provide scene context knowledge, which can be exploited e.g. as size constraint in object detection [15]. Distances of objects residing on the ground surface can be computed without depth measurement. Motion of objects is usually restricted along the surface, which can be useful during state estimation. Vertical surfaces restrict the moveable area, for the system itself, but also for all other dynamic objects within the scene. They are useful orientation clues and can provide rough directions of possible movement [26].
Because of its strong contextual information, the extraction of geometric scene knowledge has been subject of diverse work. It has been formulated as recognition problem based on texture [40] and physical plausibility [36], as optimization problem [4] jointly reasoning about the most likely configuration of vanishing points and horizon, or by geometric reasoning e.g. on line segments [50]. However, the computational complexity of these single image methods precludes a real-time application on sequences of images.
When depth measurements are available as in this work, random model sampling or region growing using local consistency criteria like surface normals are two ways to directly achieve a scene segmentation that covers large background surfaces. However, the combination of short-baseline stereo, wide angle lenses and large measurement distances in urban scenarios

poses challenging conditions. Additionally, methods need to be robust to occlusion caused by other traffic participants and static infrastructure. New measurements must update the existing knowledge about the scene in a consistent way.
Regarding the actual representation of the scene geometry, different geometric models come into question. The model complexity should be chosen depending on different factors such as the environment of operation, the quality of available measurements, the required accuracy, the range of interest and, of course, the intended usage. In inner-urban environments with a range of interest of around 20 meters, the assumption of a flat world can be justified. Modelled as a planar surface with a geometric plane, this is the most abstract, compact and computational most efficient representation. We adopt this representation also for vertical structures to meet the requirements of a very lightweight geometric background model. It consists of a common ground plane and a variable number of vertically aligned planes that represent building facades, fences and alike.
These plane models are fitted in disparity depth data and tracked over time using the estimated egomotion (Section 3.1.1). On unconstrained platforms without influence on camera orientation, surfaces are often subject to heavy occlusion and invisibility. To support model tracking in these situations, the geometric feature of vanishing points is introduced in Section 3.1.2. Plane estimation and vanishing points are combined in Section 3.1.3 to form the background model. It is then extended by the special background construct of stairways in Section 3.1.4. The estimated scene model provides strong clues for orientation. In Section 3.1.5 we discuss how the observed background structure can be fed back into the egomotion estimation in order to mitigate the drift of unreferenced odometry.

3.1.1 Feature 1: Geometric Planes

A plane in camera space can be written as

$$\mathbf{n}_x X + \mathbf{n}_y Y + \mathbf{n}_z Z + d = 0 \tag{3.1}$$

with the homogeneous plane representation $\pi = (\mathbf{n}, d)^{\mathrm{T}}$, where $\mathbf{n}$ is the plane normal vector fixed in the camera origin and $\frac{d}{||\mathbf{n}||}$ the distance to the plane. All points $\mathbf{P} = (X, Y, Z, 1)^{\mathrm{T}}$ satisfying $|\pi^{\mathrm{T}}\mathbf{P}| < \epsilon$ can be considered lying in the plane, where ϵ sets a margin around the plane.
There are different options to measure planes from image evidence. In monoscopic camera setups a popular way is to estimate the homography induced by the planar surface. Four feature point correspondences on the plane between two views are required to calculate the homography using the DLT algorithm [86]. Its parameters can then be tracked with particle [56] or Kalman filters [13]. If the transformation between the views is known, the homography can be decomposed uniquely into the plane parameters [95]. In binocular camera setups models can be fitted using the depth information, or directly by minimizing the photometric error between both views [16]. When the camera is aligned to the ground with zero roll angle, the row wise depth image histogram (also known as v-disparity) depicts flat surfaces as lines which are easily extracted by line fitting algorithms [48]. While a zero roll assumption might apply in vehicle mounted camera setups, it is usually heavily violated for free moving cameras and would require a roll compensation beforehand [113]. Therefore, we adopt the classical, less constrained method using general least squares fitting in 3D or depth data.

3.1.1.1 Measuring Planes in Euclidean Space

Three non-collinear points P_1, P_2, P_3 in Cartesian coordinates generate a unique plane

$$\mathbf{n} = \frac{(P_2 - P_1) \times (P_3 - P_1)}{||(P_2 - P_1) \times (P_3 - P_1)||} \tag{3.2}$$

$$d = -P_1\mathbf{n}^{\mathrm{T}} \tag{3.3}$$

With more than three points the problem is overdetermined. A solution is to extract the normal vector as the last principal component of the point set. The plane distance d is then calculated following (3.3) with P_1 set to the mean of all points.

In practice, we want to approximate a given point set with a number of planes. This involves two steps. First, partitioning the data into groups of points which belong to the same plane, and secondly, fitting the optimal plane to each group of points. Data partitioning can be solved following the RANSAC scheme by random plane hypothesis sampling. A random minimal set of three points is chosen to span a plane, which is evaluated by measuring the support of all remaining points. This measurement represents the probability of a point belonging to the plane and is in its simplest form 1 if the point lies within a plane distance ϵ and 0 otherwise.

Following Section 2.1, points reconstructed from binocular images are subject to position uncertainty depending on their measured $uv\delta$ image coordinates. A statistical correct measure accounts for this uncertainty in the point to plane distance function. This is primarily important in the data partitioning step to avoid discarding points in large distance with accordingly large uncertainties. Furthermore, the uncertainty of points should be considered when fitting the model in a least squares sense. An approximation for such distance function consists in employing the linearly approximated point covariance (2.3) to derive the Mahalanobis point to plane distance [79]. However, the covariance has to be explicitly computed for each image point. This computational expense can be avoided when measurements are taken in the image space instead.

Measurement errors in $uv\delta$ image space are distributed close to a normal distribution [84]. This property makes it advantageous to measure planes in image space. However, we are usually interested in their properties in camera space (e.g. their orientation with respect to the camera or their orientation with respect to each other). When measuring planes by means of least squares optimization, the objective function will calculate a difference in image space, while any optimization constraints will be given in camera space. When applying camera transformations, these are related to camera space while plane parameters are measured in image space. Hence, we will need to use both representations alongside each other. The transformation from camera space to image space is plane preserving. Thus, it is possible to use the same linear tools to measure planes in image space and transform their parameters to camera space. On the other hand, the transformation is not angle-preserving, orthogonal planes in camera space are not mapped to orthogonal planes in image space.

3.1.1.2 Measuring Planes in Image Space

We represent a plane in $uv\delta$ image space as

$$\alpha \cdot u + \beta \cdot v + \gamma + \delta(u, v) = 0 \tag{3.4}$$

Just as in Euclidean camera space, planes can be found using the RANSAC scheme by repeatedly sampling planes through 3 random points. A plane is evaluated by counting the number of support points with point-to-plane distance $|\alpha u + \beta v + \gamma - \delta|$ smaller than a disparity margin ϵ.
We want to minimize the cost function

$$\min_{\alpha,\beta,\gamma} \sum_{i=1}^{N} (\alpha \cdot u_i + \beta \cdot v_i + \gamma + \delta_i)^2, \tag{3.5}$$

which leads to the linear equation system

$$\underbrace{\begin{pmatrix} \sum u_i^2 & \sum u_i v_i & \sum u_i \\ \sum u_i v_i & \sum v_i^2 & \sum v_i \\ \sum u_i & \sum v_i & N \end{pmatrix}}_{H^{\mathrm{T}}H} \cdot \begin{pmatrix} \alpha \\ \beta \\ \gamma \end{pmatrix} = \underbrace{\begin{pmatrix} -\sum u_i \delta_i \\ -\sum v_i \delta_i \\ -\sum \delta_i \end{pmatrix}}_{H^{\mathrm{T}}\mathbf{y}}. \tag{3.6}$$

This is equal to the ordinary least-squares solution with measurement matrix $H = [u_{1..N}\ v_{1..N}\ 1]$ and the corresponding disparity measurements $\mathbf{y}$, but more efficient to compute in terms of memory consumption. Note, that this does not minimize an orthogonal point to plane distance as in camera space. The image coordinates (u, v) can be assumed error-free and are independent variables. To allow for imprecise data partitioning of plane and non-plane points and further increase the estimation accuracy, we iteratively alternate between plane support point selection and parameter optimization for a few times.

Transforming u, v and δ in (3.4) according to $F(\cdot)$ (2.1, page 10) leads to the $uv\delta$ plane to Euclidean plane parameter transformation

$$\begin{pmatrix} \mathbf{n}_x \\ \mathbf{n}_y \\ \mathbf{n}_z \\ d \end{pmatrix} \propto \begin{pmatrix} \alpha f \\ \beta f \\ \alpha c_u + \beta c_v + \gamma \\ Bf \end{pmatrix} \tag{3.7}$$

and vice versa

$$\begin{pmatrix} \alpha \\ \beta \\ \gamma \end{pmatrix} = -\frac{B}{d} \begin{pmatrix} \mathbf{n}_x \\ \mathbf{n}_y \\ \mathbf{n}_z f - \mathbf{n}_x c_u - \mathbf{n}_y c_v \end{pmatrix}. \tag{3.8}$$

3.1.1.3 Tracking Planes by Optimization

In theory, the iterative least-squares procedure from the previous section can be applied to constantly remeasure a plane over time from one camera pose to the next. The parameter estimate from the last camera pose provides the starting point for support point selection with subsequent iterative optimization in the current image. In practice however, convergence to the correct plane parameters depends strongly on the outlier ratio of the selected support points. The optimization only succeeds when camera motion between the camera poses is small enough. In certain setups this can be assumed [16, 57], but in general this cannot be guaranteed in case of unconstrained cameras. If an estimate of the camera motion between the camera poses is available, it can be applied to predict the plane parameters to initialize the least-squares refinement:

Let $T_k = \begin{pmatrix} R & \mathbf{t} \\ \mathbf{0}^{\mathrm{T}} & 1 \end{pmatrix}$ be the estimated transformation between camera poses at time $k-1$ and k. For a point $\mathbf{P}$, we have $\mathbf{P}_k = T_k \mathbf{P}_{k-1}$. For the transformation of a plane $\pi = \left(\mathbf{n}, d\right)^{\mathrm{T}}$ this results in

$$
\begin{aligned}
& & \pi_k^{\mathrm{T}} \mathbf{P}_k &= \pi_{k-1}^{\mathrm{T}} \mathbf{P}_{k-1} && (3.9)\\
\Leftrightarrow & & \pi_k^{\mathrm{T}} (T_k \mathbf{P}_{k-1}) &= \pi_{k-1}^{\mathrm{T}} \mathbf{P}_{k-1} && (3.10)\\
\Leftrightarrow & & \pi_k^{\mathrm{T}} &= \pi_{k-1}^{\mathrm{T}} T_k^{-1} && (3.11)\\
\Leftrightarrow & & \pi_k &= \left(\pi_{k-1}^{\mathrm{T}} T_k^{-1}\right)^{\mathrm{T}} = \left(T_k^{-1}\right)^{\mathrm{T}} \pi_{k-1}, && (3.12)
\end{aligned}
$$

with

$$
\left(T_k^{-1}\right)^{\mathrm{T}} = \begin{pmatrix} R^{\mathrm{T}} & -R^{\mathrm{T}}\mathbf{t} \\ \mathbf{0}^{\mathrm{T}} & 1 \end{pmatrix}^{\mathrm{T}} = \begin{pmatrix} R & \mathbf{0} \\ -\mathbf{t}^{\mathrm{T}} R & 1 \end{pmatrix} . \tag{3.13}
$$

The predicted parameters π_k allow for a proper selection of support points and are a sufficient initialization for parameter optimization. Such prediction and optimization scheme enables to track a plane under arbitrary camera movements, provided that the plane is visible and can be measured. As obvious this condition is, as often is it violated in real scenarios with uncontrolled viewing direction. Reasons are twofold. Objects and scene clutter can heavily occlude the plane up to total occlusion. Or, the camera might be oriented in a way that the plane is temporarily not visible. Both conditions occur frequently in inner-urban scenes and quickly lead to loss of tracking if not explicitly treated. Short periods of full occlusion can be handled by predicting the parameters according to (3.12), but even small drift in the estimated egomotion quickly leads to a large divergence from the true parameters with consequent loss of tracking. To deal with situations like these, robust estimation techniques are required. In Section 3.1.3.1 we will introduce a filter approach to this end, which is based on the additional geometric feature of vanishing points.

3.1.2 Feature 2: Vanishing Directions

Man-made scenes exhibit clear structures. On a low level they are composed of many orthogonal and parallel structures, for instance created by brickstones, the windows and doors in building facades, fences and many more. On a higher level, these low level structures compose more complex structures like buildings, which in turn compose streets, junctions, and so on. On all levels a high degree of alignment and parallelism is a common feature of these environments. When the scene is projected to the image space, the so called vanishing points emerge from structure which is parallel in the three-dimensional scene. A vanishing point is the projection of a point positioned infinitely far in the direction of parallel scene structure. In a calibrated camera a direction vector can be converted into an image point and vice versa, and *vanishing point* and *vanishing direction* become interchangeable names for the same property.

Vanishing directions have a few interesting properties in the context of scene understanding. Foremost, they are a property of the scene and, as such, provide a scene referenced measurement. From two known vanishing points, the camera orientation with respect to the scene can be completely recovered. Similar to a compass reading, a single vanishing point can provide a reference during egopose estimation, as will be later shown in Section 3.1.5.

Vanishing directions are obtained from a monoscopic image, and yet provide valuable spatial information about the scene. Especially the orientation of structure becomes directly measurable. Man-made flat surfaces are usually aligned with two vanishing directions. One pointing vertical corresponding to the direction of gravity, the other pointing towards the horizon. The direction orthogonal to both corresponds to the surface normal. If all vanishing points of a scene are known, large parts of the projected image can be explained in terms of orientation and alignment to each other. The orientation of an urban street canyon is contained in a single vanishing direction measurement, and the relative orientation of two streets is simply the relative orientation of their corresponding vanishing directions.

A direction measurement is independent of the observer's position. The environment does not change when the camera moves or rotates. Wherever vanishing points are measured within the local scene, their relative orientations do not change. The measurement is invariant under translational viewpoint change, only a camera rotation affects the measurement. Measuring scene vanishing directions from two different viewpoints allows to recover the orientation between these viewpoints. However, nothing can be deducted regarding the viewpoint positions. Position invariance implies that all measurements taken to estimate vanishing directions are ambiguous in terms of their camera distance. The scale of the scene remains unobservable. In this section an algorithm is proposed that detects vanishing points in sequences of images and models them consistently over time. The vanishing directions provide a powerful feature that will facilitate the recognition and interpretation of the scene background.

3.1.2.1 Measuring Vanishing Directions

Various methods have been developed to estimate vanishing points from a single image. In most recent approaches the input images are abstracted to line or edge segments. It has been shown that operating on geometric edge primitives does not necessarily come at the cost of accuracy when compared to direct estimation methods using e.g. continuous image gradients [21].

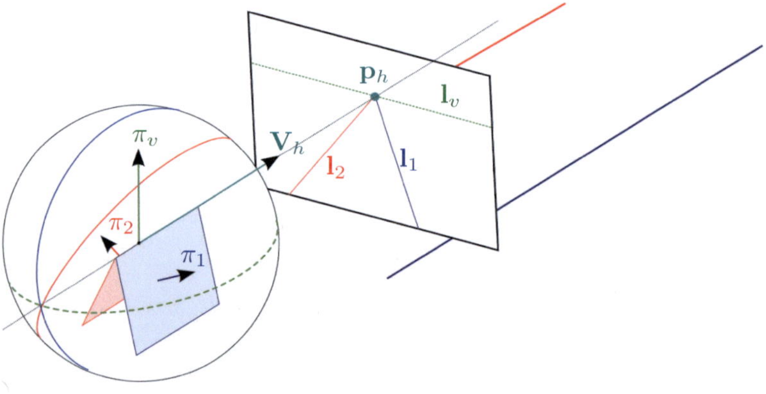

Figure 3.2: Geometric relations of lines in image space and their vanishing directions. Lines $\mathbf{l}_1$ and $\mathbf{l}_2$ intersect in the vanishing point $\mathbf{p}_h$ (vanishing direction $\mathbf{V}_h$). The horizontal plane π_v projects to the image plane as vanishing line $\mathbf{l}_v$, which corresponds to the horizon.

When aiming at real-time applications this massive data reduction is hence easily justified. Basic geometric relations allow to detect vanishing directions from line or edge segment measurements.

A line in image space can be described as $au + bv + ch = 0$ with an implicit line representation $\mathbf{l} = (a, b, c)$ and homogeneous image coordinates (u, v, h). Two image points $\mathbf{p} = (p_1, p_2, p_3)$ and $\mathbf{q} = (q_1, q_2, q_3)$ lie on the line $\mathbf{l} = \mathbf{p} \times \mathbf{q}$. Due to the principle of duality, points and lines are interchangeable in projective geometry. The intersection point $\mathbf{k}$ of two lines $\mathbf{m}$ and $\mathbf{n}$ is given by $\mathbf{k} = \mathbf{m} \times \mathbf{n}$.

In a calibrated camera with intrinsic camera matrix K, a line $\mathbf{l}$ in image space is the projection of an (interpretation) plane π through the camera origin with $\pi = K^{\mathrm{T}}\mathbf{l}$ (Figure 3.2). The intersection of two such planes, spanned by two lines $\mathbf{l}_1$ and $\mathbf{l}_2$, yields a direction vector $\mathbf{V}_h$. It intersects the image plane in $\mathbf{p}_h = K\mathbf{V}_h$ with image coordinates $(\frac{p_1}{p_3}, \frac{p_2}{p_3})$, which is coincident with the intersection of $\mathbf{l}_1$ and $\mathbf{l}_2$.

The direction $\mathbf{V}_h$ is called a vanishing direction, under the assumption that the line observations $\mathbf{l}_1$ and $\mathbf{l}_2$ were generated by scene structure which is parallel in the 3D space. When the intrinsic camera parameters K are known, each vanishing direction can be transformed into the corresponding image projection $\mathbf{p}_h$ called vanishing point, and vice versa. Both terms are used interchangeably in the remainder. The intersection of each pair of line observations creates a potential vanishing point. However, due to the projection ambiguity it is impossible to detect a real vanishing point from only two line observations. Robustly recognizing vanishing directions requires to analyse the scene in its entirety.

The problem of detecting a scene's vanishing directions is a problem of multi model fitting and can be treated with different strategies, which all need to deal with a few common sub-problems. First, a method to hypothesize potential vanishing points is required. To evaluate these hypotheses, an assignment needs to be found between line segments and vanishing point hypotheses. This requires a consistency measurement which quantifies the support of a line to a vanishing point. With the assigned, or grouped line segments, the vanishing directions need to be estimated that explain the assigned observations in an optimal way.

In his early work of [5] Barnard suggested a Hough transform to find vanishing point candidates, in which the Gaussian sphere is used as voting space. The interpretation planes of line observations intersect the sphere in great circles and create modes on the sphere which correspond to the vanishing directions (Figure 3.2). A line observation $\mathbf{l}$ is assigned to a vanishing direction $\mathbf{V}$, if $\mathbf{V}$ lies within the interpretation plane π_l according to their inner product $\langle \mathbf{V}, \pi_l \rangle$.
Alternatively, the analysis can be carried out in the (unbound) image space directly by evaluating a pixel error between line and vanishing point. Various variants exist, the most commonly used consistency measurements were recently evaluated in [93]. Their deficiencies in terms of ignoring edge length or being variant to image position are often a trade-off between optimality and simple and fast computability. Clearly, consistency measuring is also possible directly using the continuous space of image gradients [17, 78, 59], with the drawback of high computational costs.
A consistency measure allows the immediate application of unsupervised data analysis, like k-means, multi-model RANSAC or any hierarchical clustering algorithm [87] to group the edges. These methods have the advantage of operating in continuous parameter space, which obviates the difficult choice of parameter discretization as required in the classic implementation of Hough space methods. After clustering, optimal vanishing directions are the result of non-linear refinement. To avoid hard assignments between edge observations and vanishing directions, the expectation maximization framework is often utilized in this context [2, 45, 78].
Many recent methods make use of prior scene knowledge. Man-made environments usually follow a clear alignment with the direction of gravity, building facades are flat and meet each other in distinct angles. In the extreme case, structures are aligned to each other in an orthogonal fashion. These scenes are composed of three orthogonal vanishing directions and are referred to as a *Manhattan world*. In these settings, where the arrangement of vanishing directions is known, all directions can be estimated simultaneously [17, 62, 6, 94]. Unfortunately, the Manhattan world assumption is often violated and renders it necessary to also estimate the scene geometry, at least partly. A relaxation was proposed in [78] with the so called *Atlanta world*.

Figure 3.3: Vanishing directions detected independently from 10 slightly different viewpoints in a sequence of images. Most detections are correct in terms of the scene layout, but show varying amounts of noise. Background image: Google Earth, © 2009 GeoBasis-DE/BKG.

Here, the Manhattan frame is extended by arbitrarily many horizontal vanishing directions, which are orthogonal to a common vertical axis that corresponds to the direction of gravity. Such representation can depict the typical layout of the majority of man-made environments, which are usually aligned with gravity. It is the representation that we will also adopt in this work.

Independent of the measurement and model assumptions, the estimates of all methods are afflicted with noise which is propagated from sensor level to image gradients and edge segment measurements. This noise inevitably affects the accuracy of an estimated vanishing point. Especially scene parts located far away from the sensor become problematic, since they are usually depicted only with short and noisy edge segments. The variance in initialized vanishing directions is large in these cases. Figure 3.3 shows the situation. Here, vanishing directions are initialized using the state of the art approach of Tardif [87] independently for all images of a short sequence. Using the ground truth camera orientation from an inertial measurement unit (IMU) the four most certain vanishing points from each position are registered in the IMU reference frame. None of these initializations is wrong in terms of scene layout, but their estimates are subject to obvious uncertainties. The forward direction is initialized with small variance, the parking cars create a very noisy vanishing point roughly orthogonal to it. Here, we are interested

in the intersecting street, which can be recognised in a blurred cluster. How should this direction be initialized with respect to the frontal direction, using the sequence of images?
The evidence of each single viewpoint is not very strong, but we can expect the detections from many slightly different viewpoints to accumulate around the real directions. The center of these clusters will statistically be a better estimate than each single view detection. A strategy should thus be, to accumulate measurements and defer the initialization until enough evidence is available. This is hindered by the fact, that the orientation between the different camera poses of the sequence is usually not known. To accumulate vanishing point detections, we have to know the camera orientation with respect to the scene. However, we can estimate the camera orientation only robustly, if the scene layout is known. Tracking an assembly of known vanishing directions (termed *vanishing direction model* in the remainder) and tracking camera orientation become interchangeable problems in this case.
Tracking vanishing points can been understood as a constrained multi object tracking problem, and can thus be handled with the same tools. Following the tracking by detection scheme, vanishing points would be detected independently in consecutive images and associated to form a track. While this is feasible, when the model structure is known (e.g. Manhattan [22]), it becomes very sensitive when the model has to be estimated along. Apart from that, valuable information from the past remains unconsidered. History can be incorporated by casting the problem into a min flow cost optimization of a graph over a batch of images, as recently proposed in [46]. While achieving convincing results, such kind of batch processing is computationally expensive. To retain real-time capabilities, a sequential update strategy would be desirable, which considers the measurement history, but keeps track of the model using only the current, new measurements.
Such strategies have been implemented for instance in the frameworks of Bayesian inference [17, 59] or expectation maximization [78]. Vanishing directions are tracked by optimizing their parameters based on their last known estimates, instead of detecting and associating them independently. However, it is assumed here again that the model itself does not change during the sequence. Recognizing a change in the model (e.g. due to an emerging

side street), or even recognizing the loss of tracking is difficult since the optimization might converge in local minima. In this case an independent detector can provide a solution.
Our approach combines both methodologies. We will track the model of vanishing directions by continuously optimizing its parameters to adapt it to the changing camera orientation, while we simultaneously adapt it to the changes in the observed scene. To judge about the correctness of the model and its number of parameters, we propose an algorithm that detects vanishing directions independently. By accumulating these detections over time we can defer decisions about model changes until enough evidence has accumulated. That way we make use of the long history of past detections.

3.1.2.2 Tracking Multiple Vanishing Directions

Our approach is based on edge segments. We experiment with two different edge extraction methods. An implementation based on a Canny edge detector taken from [87], and the recent approach of EDLines [1]. Both provide edge segments fitted with a line model $\mathbf{e}$, which set up the edge list $\mathbb{E}$.
We measure the consistency between a vanishing direction V and an edge segment $\mathbf{e}$ in image space as the orthogonal distance between one of the line endpoints e_1 and the line $\mathbf{l}$, that connects the vanishing point $\mathbf{v}$ with the edge centroid e_c (Figure 3.4). The distance D is calculated as

$$D(\mathbf{v}, \mathbf{e}) = \text{dist}(e_1, \mathbf{l}) = \frac{\langle e_1, \mathbf{l} \rangle}{\sqrt{l_1^2 + l_2^2}}, \text{ with } \mathbf{l} = e_c \times \mathbf{v}. \tag{3.14}$$

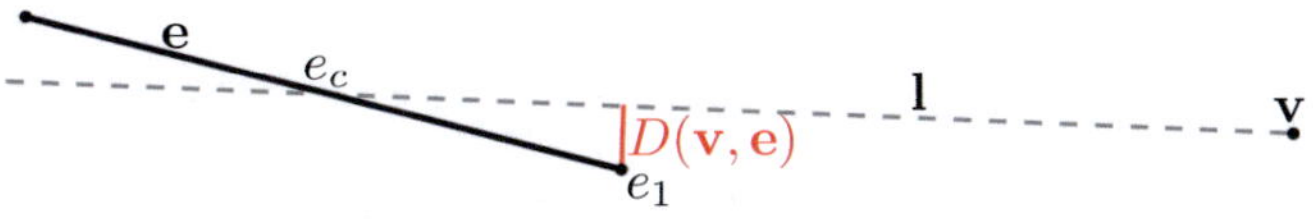

Figure 3.4: Error D between edgelet $\mathbf{e}$ and vanishing point $\mathbf{v}$.

By evaluating $D < \epsilon$ for a set of edges $\mathbb{E}$ we can select the support edges $\mathbb{E}_V$ for a given vanishing direction $V(\theta, \varphi)$ and a distance threshold ϵ. The vanishing direction that maximizes the consistency with $\mathbb{E}_V$ is found by minimizing

$$V^+ = \min_{\theta,\varphi} \sum_{\mathbf{e}\in\mathbb{E}_V} D(\, K\, V(\theta,\varphi), \mathbf{e}) \tag{3.15}$$

where V is defined by the minimal two spherical coordinates θ and φ. The intrinsic camera matrix K projects the vanishing direction V to its vanishing point on the image plane.

Depending on the intensity of camera motion this minimization can be already sufficient to track a vanishing direction over sequences of images. The optimization usually converges after 3 to 4 iterations between inlier selection and minimization. Problems can be caused by spurious inlier selection. Due to the direction ambiguity of line observations, edges might be selected as inlier, although they belong to completely different vanishing directions.

In practice, scenes are composed of multiple vanishing directions, which should all be tracked simultaneously. The previous inlier selection is now modified into an assignment between edges and vanishing directions. An obvious solution to this is nearest-neighbour assignment using $D(\cdot)$. Related work has also treated this task as soft-classification using expectation maximization [2, 45, 78]. The idea is to iterate between a weighted assignment in the expectation step, and the optimization of all vanishing directions in the maximization step using the current weight. In our experiments this scheme performed most robustly, when ambiguously assigned edges with mixed weights were completely disregarded in the maximization step. This lets us opt for an efficient two step solution. In a first step, inlier edges are found for each vanishing direction independently. Afterwards, all edges are rejected which were assigned to more than one direction. Maybe this reflects reality, where the assignment is strictly distinct and an edge cannot belong to two different vanishing directions.

Having clustered the edges, we can proceed to optimize each vanishing direction with the respective inlier edges using (3.15). Since vanishing directions are fixed with respect to each other, the optimization would need to be complemented with according constraints (e.g. to enforce a Manhattan world model: $\langle V_1, V_2\rangle = \langle V_2, V_3\rangle = \langle V_1, V_3\rangle = 0$). A more intuitive and efficient way is to rotate the camera, instead of rotating the single vanishing directions. We seek the spatial rotation R that needs to be applied to all vanishing directions in order to minimize their cost functions. This is expressed as

$$\min_{\theta,\phi,\psi} \sum_{V_n} \sum_{\mathbf{e}\in\mathbb{E}_{V_n}} D(\ K\ R(\theta,\phi,\psi)\ V_n, \mathbf{e}). \tag{3.16}$$

The updated vanishing directions are then obtained as

$$V_n^+ = R(\theta,\phi,\psi)\ V_n. \tag{3.17}$$

This has two big advantages: We have to optimize 3 rotation parameters for arbitrary many vanishing directions instead of $2\ n$ parameters for n directions in an independent optimization. Furthermore, each edge contributes to each direction. This leads to more observations to estimate fewer parameters. Consequently, the estimation is much more robust. The prerequisite is, that the arrangement of vanishing directions correctly models the scene layout. In inner-urban scenarios, the street layout changes frequently during traversal, mainly due to side streets that emerge and disappear. We are able to track the camera orientation jointly with the given vanishing directions of a scene. For practical applicability, we need to adapt the model to the environment over time.

3.1.2.3 Vanishing Direction Model

Our scene model is based on the Atlanta world of [78]. Here, vanishing directions are arranged orthogonally to a common vertical axis, which corresponds to the direction of gravity. This model can represent the majority of inner urban scenes, while it offers great simplifications for efficient estimation.
All horizontal vanishing directions lie in a plane $\boldsymbol{\pi}_v$ through the camera origin (Figure 3.2). The projection of $\boldsymbol{\pi}_v$ to the image plane is the vanishing line $\mathbf{l}_v$ which corresponds to the horizon. All vanishing directions coherent with our model project to vanishing points which lie on $\mathbf{l}_v$.
Under the assumption that an edge segment $\mathbf{e}$ belongs to a horizontal vanishing direction which lies within $\boldsymbol{\pi}_v$, it can be transformed into a 3D direction.

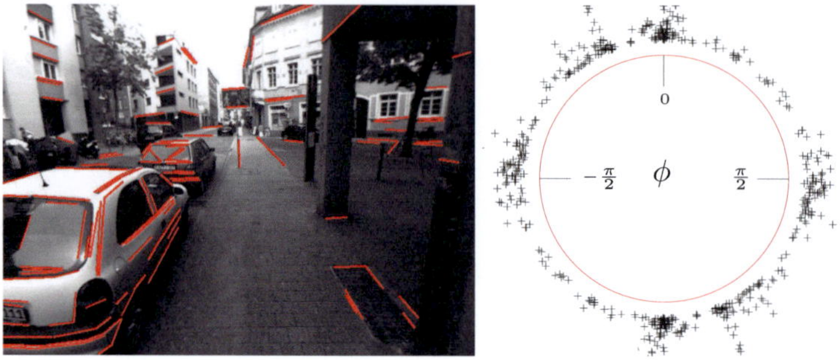

Figure 3.5: All hypotheses for horizontal vanishing directions obtained from the red edge segments. A zero azimuth angle ϕ corresponds to the frontal direction and edge length is encoded with radius.

The potential vanishing point in image space is given as the intersection of the edge with the vanishing line $\mathbf{l}_v \times \mathbf{e}$, or, its direction E in Euclidean space as

$$\begin{aligned} \boldsymbol{\pi}_e &= K^T \mathbf{e} \\ E &= \boldsymbol{\pi}_e \times \boldsymbol{\pi}_v. \end{aligned} \tag{3.18}$$

A horizontal vanishing direction can now be represented by a single parameter ϕ as the angle to an arbitrarily chosen frontal direction F lying in $\boldsymbol{\pi}_V$. Using (3.18), each edge generates one datapoint $\phi_e = \langle F, E \rangle$ in an accumulator $\mathbb{A}$. Figure 3.5 shows the accumulator for an exemplary single frame. The accumulator contains all edges that do not belong to the vertical direction according to the data association in Section 3.1.2.2. Intuitively, modes in this accumulator entail the horizontal vanishing directions of the scene. Clusters can be recognised at an azimuth angle ϕ of 0° meeting the frontal vanishing direction, at around -30°, corresponding to the side street, and at 90° mainly caused by the parking cars.

A model of at least two vanishing directions is sufficient to track the camera orientation from frame to frame using the method presented in Section 3.1.2.2. The current orientation of the camera R_{W_k} to an initial reference frame W updates according to $R_{W_k} = R(\theta, \phi, \psi) R_{W_{k-1}}$ from frame to frame, with θ, ϕ, ψ being the solution of (3.16), and k denoting the current frame.

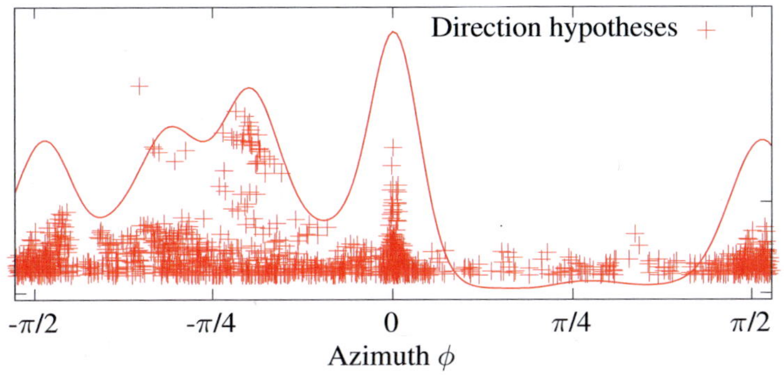

Figure 3.6: Vanishing direction hypotheses generated from all non-vertical edges, accumulated in a reference frame over 10 images. The data value corresponds to the edge segment length.

Knowing R_{W_k} we can accumulate the direction hypotheses in W over an arbitrarily long history of frames. Evidence for new vanishing directions moving into the view accumulates over time, while it disappears in the same manner when vanishing directions move out of view. Angles at which measurements pile up are an indicator for the prominent scene directions. We will use this to decide about adding or removing horizontal directions from the model. The correct model in turn allows to track the camera orientation.

Accumulator Analysis Figure 3.6 shows the accumulator for a history length of 10 frames. We want to use this feature space for two purposes: On one hand, decide whether a certain vanishing direction needs to be added to the model. On the other hand, decide whether existing vanishing directions should be kept or be discarded. In both cases, the decisive feature is the amount of support from the accumulator. Measuring support for a given direction is straight forward, the simplest solution would sum up all data values within a margin around the respective azimuth angle. By contrast, finding new directions requires to analyse the whole accumulator for clusters of data. Computational cost puts a narrow limit on the amount of data and thus the size of history length here.

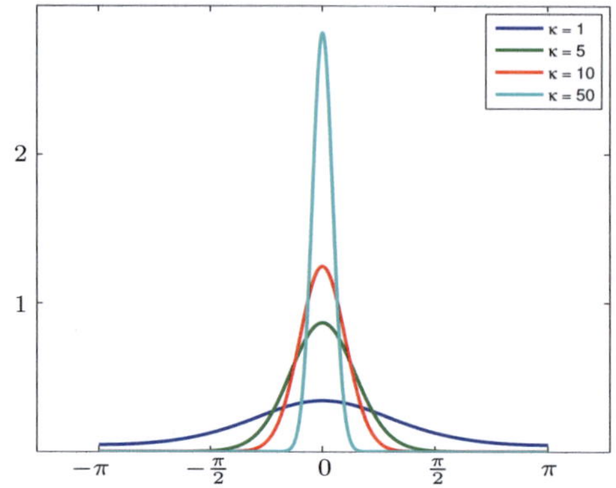

Figure 3.7: The von Mises distribution for $\mu = 0$ and different concentration parameters κ. It resembles a normal distribution wrapped around a circle.

Our method is based on non-parametric feature space analysis using a kernel density estimator

$$\hat{P}_h(x) = \frac{1}{hn} f_h(x) = \frac{1}{hn} \sum_{i=1}^{n} w_i K\left(\frac{x - x_i}{h}\right), \tag{3.19}$$

which estimates the probability density function P_h of data points x_i with a kernel function $K(\cdot)$. A bandwidth parameter h controls the degree of smoothing. The individual data points x_i might be weighted with a normalized weight w_i which satisfies $\sum_{i=1}^{n} w_i = 1$. The kernel $K(\cdot)$ should reflect the circularity of our data. We use the von-Mises distribution

$$f_{VM}(x \mid \mu, \kappa) = C(\kappa) e^{\kappa \cos(x-\mu)},$$

which approximates the normal distribution wrapped around a unit circle. μ is the center of the distribution and κ is a concentration parameter which controls the spread of the distribution (Figure 3.7). $C(\kappa) = (2\pi I_0(\kappa))^{-1}$ normalizes the distribution, $I_0(\cdot)$ is the modified Bessel function of order 0. The Kernel becomes $K(x) = C(\kappa)\exp(\kappa \cos(x))$.

The unscaled density profile $f_h(\cdot)$ is shown as a red curve in Figure 3.6 ($h = 1, \kappa = 100, w_i \propto$ length of edge). The locations of potential vanishing directions are clearly exposed as modes of the profile.
Locating the modes of such function is the objective of the mean shift algorithm [25]. The mean shift procedure is an iterative algorithm that shifts a data sample into the maximum gradient direction until it converges in a local maximum. It is attractive for the analysis also of larger feature spaces, since it does not require to estimate the complete shape of $\hat{P}_h(x)$ explicitly. Only the weighted mean of the data x_i around the shifting sample has to be estimated.
The mean shift is defined as

$$m(x) = \frac{\sum_{i=1}^{n} w_i K(\frac{x-x_i}{h}) x_i}{\sum_{i=1}^{n} w_i K(\frac{x-x_i}{h})} - x. \tag{3.20}$$

Modes $\mathbb{M}$ are located by iteratively updating a set of sample points s_i according to $s_i \leftarrow m(s_i)$ until convergence. Samples that converge into the same mode are merged in a post-processing step. The initial sample points can be chosen equally distributed over the parameter space.
For each mode $m \in \mathbb{M}$ the support $s_m = f_h(m)$ (3.19) is evaluated. The model is extended with a new vanishing direction, if s_m exceeds an initialization threshold. Evaluating the support $f_h(v_i)$ for the model's existing vanishing directions v_i allows to recognize directions that do not exist anymore.
The computational cost of the mean shift procedure depends linearly on the amount of data points x_i that have to be analysed. When looking at the input edges in Figure 3.5 it becomes obvious that many edges originate from scene clutter (here mainly cars) which should at best be neglected during the estimation. In the accumulator it causes a large amount of background noise. From a point of speed up, stability and robustness it is desirable to increase the signal to noise ratio here. True vanishing directions are supported by a whole group of edges. Instead of using each single edge to generate a data point in the accumulator, we will use groups of edges that clearly belong to the same direction. Finding such groups of edges leads back to the problem

of detecting vanishing points. However, the detection can be restricted to vanishing points that conform with the model.

Edge Clustering Very intuitively formulated, two edges are geometrically similar when they support the same vanishing point. Given a set of vanishing points, we first evaluate, which vanishing points are supported by which edge. If two edges vote for the same vanishing points, they are likely similar to each other and can be grouped into the same cluster. This is the basic idea behind the J-Linkage clustering algorithm presented in [88] and applied to the problem of vanishing point detection in [87]. For each edge $\mathbf{e}_n$ a binary feature vector called *preference set* is build that contains one entry for each vanishing point $\mathbf{v}_m$. It is 1 if $\mathbf{e}_n$ supports $\mathbf{v}_m$ according to $D(\mathbf{v}_m, \mathbf{e}_n) < \epsilon$ (3.14), and 0 otherwise. Afterwards, the edges are grouped based on the Jaccard coefficient of their preference sets $J(A, B) = \frac{|A \cap B|}{|A \cup B|}$, which is 0 for disjoint sets and 1 for identical sets. The basis for the algorithm is a set of vanishing point hypotheses $\mathbf{v}_m$. To be in accordance with our model, a hypotheses must lie on the vanishing line $\mathbf{l}_v$ (Figure 3.2). As previously discussed, each edge generates one such hypotheses (3.18). These form the input of the algorithm, the output are clusters of edges. Each cluster is fitted with the optimal vanishing direction using (3.15) and added to the accumulator.

Figure 3.8 demonstrates the difference to the direct method, in which each edge is treated as measurement in the accumulator. The results are measurements which are quantitatively less, but qualitatively representing the real scene structure in a more accurate way with less noise. Additionally, the reduced amount of data allows to accumulate measurements over a longer history of images.

3.1.2.4 Implementation Overview

The overall implementation consists of two parts: Tracking the vanishing points by continuously optimizing the extrinsic camera orientation with respect to the vanishing direction model. This succeeds, as long as the model correctly represents the scene structure. To adapt the model, we use the support $f_h(\phi)$ over directions ϕ which is provided by an independent detector. The modes of f_h tell about potential new directions. The support of current model directions reflects their correctness and currentness. We define three support levels to this end (Figure 3.8). We initialize a new direction from a mode m when $f_h(m) > T_{init}$. The new direction is initially tracked independently along with the model (3.15). If it does not diverge from the model it is eventually added as new direction, or applied as update to an existing direction, if it falls below the minimal accepted angle between model directions (15°). For each direction in the model v_i we estimate $f_h(v_i)$ and deactivate the direction if it falls below the level $T_{deactivate}$. In this state it is kept in the model but ignored during the camera orientation estimation. This is useful to allow for temporary invisibility due to changes in the viewing direction. The direction is either reactivated, when $f_h(v_i)$ raises above $T_{activate}$, or otherwise deleted from the model after a certain time. A hysteresis between $T_{deactivate}$ and $T_{activate}$ prevents that invisible directions become accidentally reactivated by scene clutter. A high-level overview of the processing sequence is given on the next page.

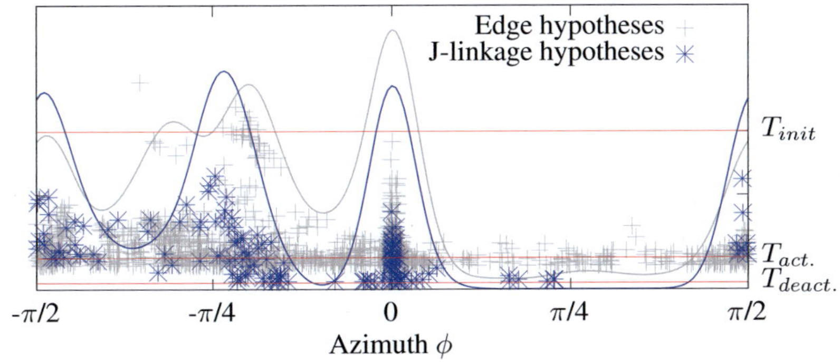

Figure 3.8: Vanishing direction hypotheses generated from clustered non-vertical edges, accumulated in the reference frame over 30 images. The data value corresponds to the cluster size. Hypotheses generated directly from edges (Figure 3.6) are underlaid for comparison. Three levels $T_{\{init,activate,deactivate\}}$ are used to decide about model modifications, see text for details.

Algorithm 1: High-level overview of the proposed method.

Initialize a minimal direction model $\mathbb{V}$ consisting of the vertical axis and the most prominent vertical direction with existing single image methods (e.g. [62, 6, 94, 87])
Initialize empty accumulator $\mathbb{A}$ and initial camera pose R_W
foreach *new image at time* k **do**
    Extract all d edges $\mathbb{E}$ from the image
    Estimate change of camera orientation R with all active vanishing directions (3.16)
    Update camera pose: $R_{W_k} = R(\theta, \phi, \psi) R_{W_{k-1}}$
    Generate d vanishing point hypotheses $\mathbb{V}_H$ (3.18)
    Run J-Linkage with $\mathbb{V}_H$ and $\mathbb{E}$ to find direction hypotheses H
    Add directions H_i to accumulator $\mathbb{A}$ as $R_{W_k} H_i$
    Delete data from $\mathbb{A}$ that is older than 30 frames
    Find modes $\mathbb{M}$ in $\mathbb{A}$ using meanshift
    foreach *mode* m *in* $\mathbb{M}$ **do**
        Calculate $f_h(m)$ (3.19)
        if $f_h(m) > T_{init}$ **then**
            Add direction of m to model $\mathbb{V}$ or
            Update existing model direction
        end
    end
    foreach *direction* v *in model* $\mathbb{V}$ **do**
        Calculate $f_h(v)$ (3.19)
        if $f_h(v) < T_{deactivate}$ **then**
            Deactivate v
            If deactivated since more than 30 frames, delete v
        end
        if $f_h(v) > T_{activate}$ **then**
            Activate v
        end
    end
end

3.1.2.5 Evaluation

The eventual goal was to measure the scene vanishing directions in local camera coordinates for each frame of a sequence. There are two aspects to evaluate: 1. How accurate is the measurement of a vanishing direction? 2. How accurately do the measurements represent the structure of the scene? Both aspects are interrelated.

The accuracy of a direction measurement can be quantified as the angular deviation between the ground truth direction and the measured direction in camera coordinates for each analysed image of the sequence. The proposed method measures vanishing directions indirectly by estimating the camera rotation between two images. Hence, the camera orientation error reflects the measurement error of the vanishing directions. We use an IMU attached to the camera to measure a ground truth frame to frame orientation delta $MR_{imu}M^{\mathrm{T}}$. This is directly comparable to the frame to frame orientation estimate R using the vanishing direction model. M denotes the extrinsic orientation between IMU and camera (Section 4.1).

Figure 3.9 shows the error distribution for R in yaw, pitch and roll angles for an indoor sequence of 240 m length. The path and exemplary measurements of this sequence are shown on pages 75 and 76. To cut out errors originating from inaccuracies in the estimated scene model, the model was set fixed to the three orthogonal directions of the building. The errors are similar in all axes with virtually zero mean error.

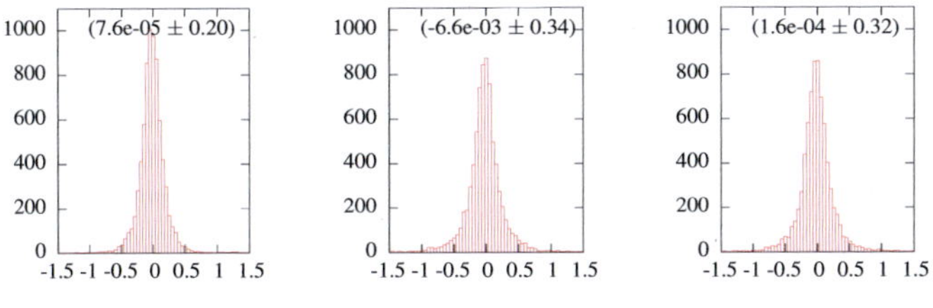

Figure 3.9: Frame to frame camera orientation errors compared to IMU ground truth in a Manhattan scene with correct model. Plots show histograms of errors in degrees divided into yaw, pitch and roll rotation, including (mean ± standard deviation).

The error in R affects the accuracy of each vanishing direction in a different way. For instance, a yaw error in R will not have any effect onto the vertical axis, but will affect all horizontal vanishing directions. Figure 3.10b shows the direction measurements mapped into the reference IMU frame, where the different noise distributions become apparent. The noticeably higher variance in the horizontal directions is explained by the fact, that the head worn camera typically rotates most around the vertical yaw axis. Inlier line segments for the horizontal directions can differ strongly between consecutive frames. Besides that, the accuracy depends on the scene itself. In sparsely built-up areas horizontal directions are often poorly supported and result in higher uncertainties. The standard deviation for the vanishing directions in this sequence amounts to around $1°$ for the azimuth angle and $0.6°$ for the elevation angle.
In the usual application the scene model is unknown and needs to be estimated along with the camera orientation. The question that arises is how accurately the model directions represent the scene structure. An indicator are the relative angles between model directions, which should conform with the real world scene. A ground truth for these angles can be determined from maps and high resolution satellite imagery. One can expect a measurable ground truth precision between $0.5°$ and $1°$, depending on the scene.

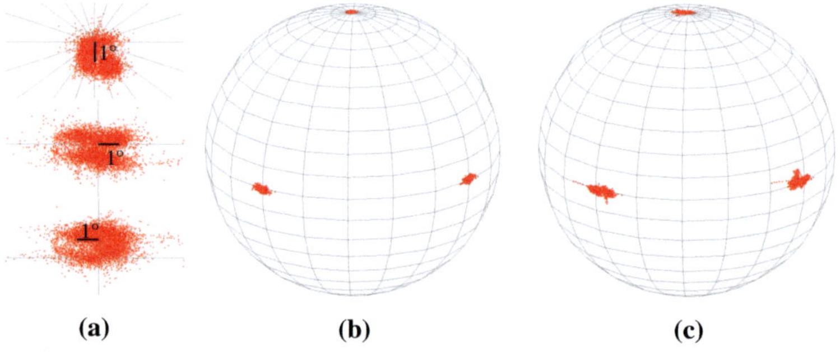

Figure 3.10: Local vanishing direction measurements mapped into the reference IMU frame. (b) shows the variance for the correct (Manhattan) model (magnification in (a)). In (c), an erroneous model was applied for estimation, which causes a higher variance.

We evaluate the estimated relative intersection angles for a dataset of inner urban scenarios. Captured using a head-worn camera (Section 4.1) the sequences consist mainly of sidewalks with turns into branching sidestreets. Figure 3.11 shows 15 such junctions and the estimated horizontal model directions. The measurement is taken shortly after the direction was added to the model, the position is indicated in the map. The average absolute error over all sequences adds up to $2.04 \pm 1.86°$. Replacing the Canny edge detector with the faster EDLines Detector reduces the accuracy and precision marginally to $2.53 \pm 2.25°$. The biggest error with $9°$ (EDlines) can be noted in scene (4). The branching street is not yet visible but already initialized only based on the ground structure. Outlier measurements on the crane in the background and the opposite building lead to the comparably large deviation. When combining both parts – model estimation and camera orientation tracking – the question arises in how far errors in the estimated scene model affect the accuracy and in general the ability of keeping track of the model. In Figure 3.10c the lateral direction of the model was disturbed by $6°$. A higher variance in the estimated directions becomes obvious, but the effect is difficult to quantify since it depends strongly on the scene and camera orientation. The optimization of R is influenced by the amount of support in the different directions. An erroneous model usually causes R to converge into a minimum, which is accurate for the strongest supported direction but rolled around this direction. In practice, the frontal direction usually has most support, followed by the vertical and the lateral direction. A wrong model is fitted into the scene with a small roll bias in this case, but camera orientation can still be tracked (albeit with increased variance).
Loss of tracking becomes apparent in identity switches of horizontal model directions. The tracking mechanism requires a minimal number of one horizontal direction. As long as the camera is in motion and is roughly oriented horizontally, this minimal configuration usually converges into a minimum after few images. After an id switch happened, a stable new configuration is automatically recovered within few frames, but the global orientation with respect to the scene gets lost. Id switches are usually not caused by slight errors in the model, but rather by a combination of strong camera motion and missing or wrongly assigned spurious edge segments in very cluttered or

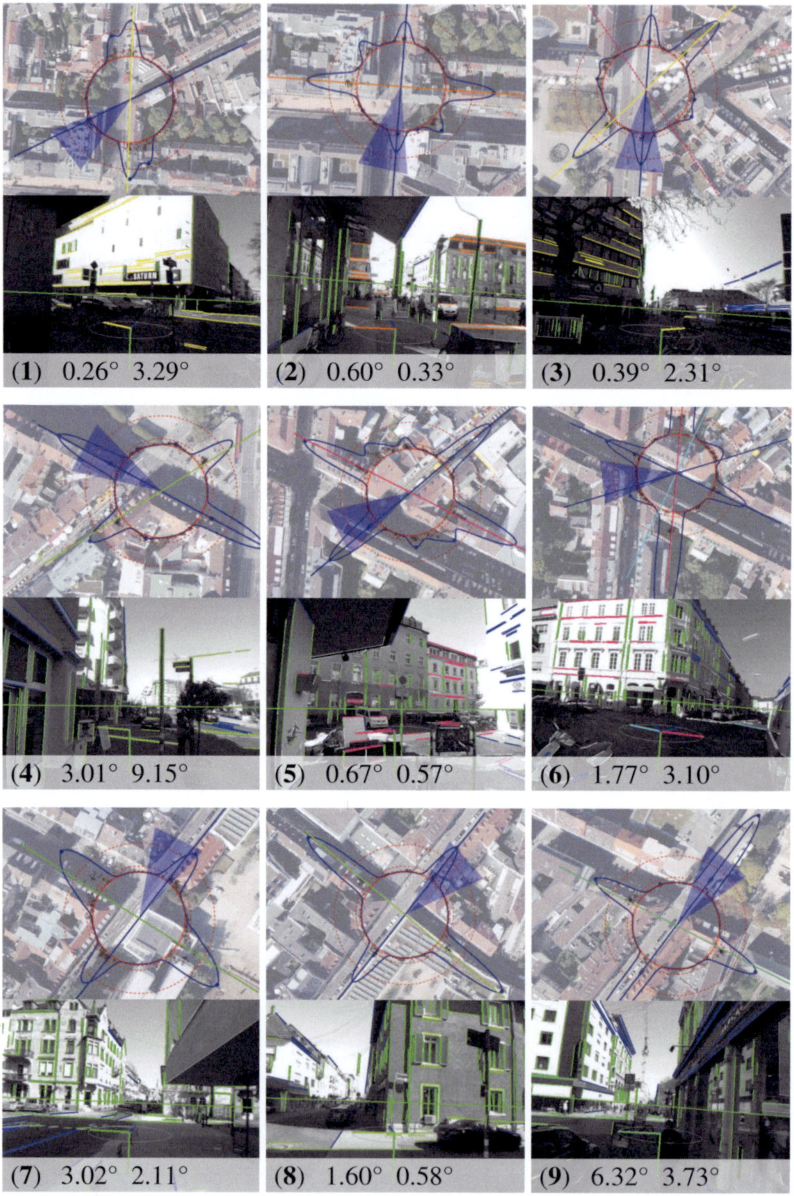
(1) 0.26° 3.29°
(2) 0.60° 0.33°
(3) 0.39° 2.31°
(4) 3.01° 9.15°
(5) 0.67° 0.57°
(6) 1.77° 3.10°
(7) 3.02° 2.11°
(8) 1.60° 0.58°
(9) 6.32° 3.73°
SATURN

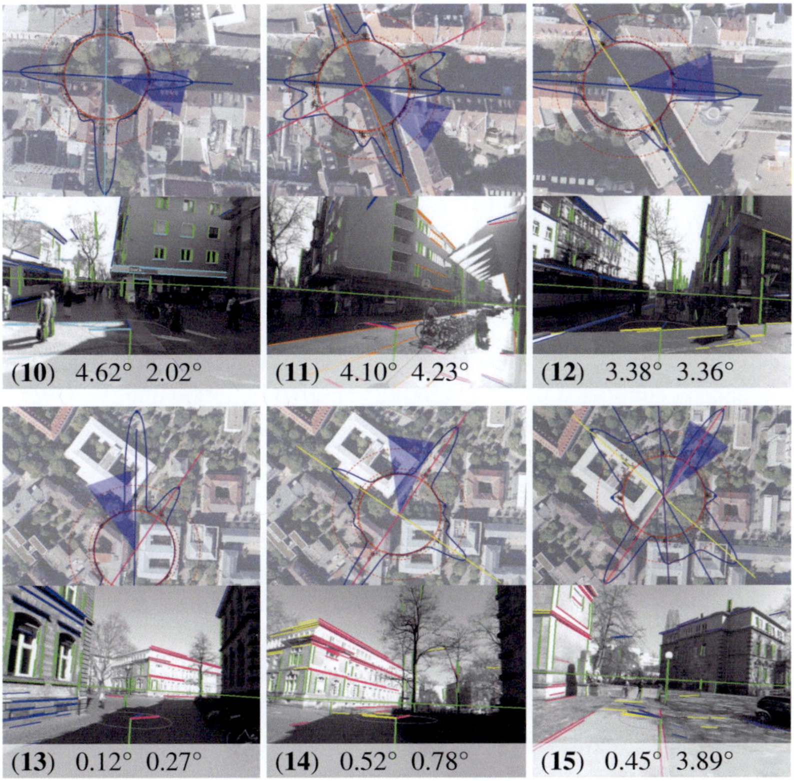

Figure 3.11: Sequences tested to analyse the accuracy of initialized directions. The support over directions is plotted as blue line, mean shift modes are indicated by dots on the line. The thresholds $T_{deactivate} < T_{activate} < T_{init}$ are plotted as dashed red circles, the model directions with colored lines, and the camera heading as blue cone. Relative errors between directions compared to ground truth are given underneath for EDLines and Canny edge segment detection. Background images: Google Earth, © 2009 GeoBasis-DE/BKG.

(a) **(b)**

Figure 3.12: (a) Three failure cases in unstructured scenes and ill-conditioned lighting conditions. (b) Loss of tracking upon entering a staircase. The vertical component flips into the horizontal plane and cannot be recovered automatically.

unstructured environments (see Figure 3.12 for three examples). The evaluated dataset contains 3.5 km of walking captured in 34000 images. Overall the correct model orientation is lost 15 times with automatic recovery. Unrecoverable tracking loss can happen when the vertical component switches into the horizontal plane. This situation is usually result of very sparse edge measurements, and occurred once in the dataset during a staircase entry (see Figure 3.12b). The orientation estimation as introduced is unfiltered. A single strong outlier estimate can cause these errors. Many failure cases could be caught by evaluating the estimated frame to frame orientation R against a reasonably chosen bound of the expected camera rotation rate. Alternatively, an independent odometric measurement can easily be used as prediction to initialize the optimization in Section 3.1.2.2. This way also very strong camera motion can be handled.

The processing time of the algorithm allows real-time operation on a wearable platform with an average runtime of around 25 ms per frame on a single CPU core. Bottlenecks are the line extraction and J-linkage stages. The latter depends on the numbers of line segments and sampled direction hypotheses, which both could be limited to reach a desired constant computation time. The mean-shift algorithm is computed for 10 equally positioned probes to locate the density modes and finishes within 1-2 ms for an accumulator size of 30 images. With these processing times it is absolutely competitive to related work that operates on image sequences. The batch processing algorithm of [46] operates in 2 s per image, the very recent Kalman filter approach of [51] with frame to frame line tracking computes 40 ms per frame. Earlier implementations that focus on single image optimization (e.g. [17, 21, 87]) finish between seconds and minutes. The proposed method is thus one of the very few methods with actual real-time applicability.

3.1.2.6 Conclusions

The aim of this work was to recover the vanishing directions that underlie a local scene. Other than in the majority of existing work, only minimal assumptions regarding the scene structure are made. The proposed method jointly estimates the parameters of a vanishing direction model and the orientation of the camera with respect to this model. To exploit the fact that moving mobile systems permanently change their viewpoint and reveal new information about the scene, the problem is treated using sequences of images as input.

The key-problem is the unknown model and especially its unknown and changing number of parameters. Every multi-instance tracking algorithm requires a decision regarding the initialization of new and closing of existing tracks. When implemented following the tracking-by-detection scheme, an additional assignment decision is required between tracks and measurements. An alternative strategy are approaches that seek the tracked instance by optimizing its last known state using correlation or likelihood maximization. Here, tracking loss is hard to recognise, and additional mechanisms are required to estimate the number of instances or model parameters. The proposed method combines both strategies. Vanishing directions are tracked by optimizing the camera orientation with respect to the model. An independent detector accumulates evidence using the history of images. This evidence over scene directions offers an elegant way to decide about required modifications to adapt the model to changes in the passing scene.

The evaluation shows a measurement precision of vanishing directions within 1° standard deviation for a given scene model. The accuracy of the estimated scene models varies around 2°. The algorithm itself is very lightweight and one of very few implementations applicable in real-time on a single core of current standard hardware.

The ability of measuring scene orientations can be an enabler for a variety of different applications. In a local view, vanishing directions provide accurate information about surface orientations. In the following Section 3.1.3.1 this is utilized to track planar surfaces even in cases where they are not directly visible. Once a line observation is assigned to a vanishing point its 3D orientation is known. This facilitates the correct reconstruction of line segments from multiple viewpoints. In Section 3.1.4 an algorithm is introduced that uses parallel lines and their orientation to detect staircases along with their geometric properties. Vanishing points can furthermore provide valuable context information during obstacle detection, since foreground objects are often aligned with the scene geometry. An example will be given in Section 3.2. In a global view, vanishing directions can be exploited to estimate the camera orientation with respect to the scene, which is a byproduct of the proposed algorithm. In combination with odometric measurements, a drift free, scene referenced position estimate can be achieved, as will be demonstrated in Section 3.1.5.

3.1.3 Scene Background Model

Two features were introduced to measure properties of the scene background: geometric plane models, which can be measured from depth data by combining random model sampling and least-squares optimization, and vanishing directions, which can be extracted from the image information. In this section, both features are combined in order to estimate a scene background model. It consists of a ground plane, complemented by a varying number of vertical planes, which represent background structures like buildings, fences or bushes. The ground plane is of particular importance, since it provides the common reference. It needs to be reliably tracked, also during ill conditioned situations where it is not visible.

Figure 3.13: Three situations in which objects and scene clutter heavily occlude the ground plane.

3.1.3.1 Tracking Occluded and Invisible Planes

As detailed earlier, tracking plane parameters can principally be done by iterative least-squares optimization. In real applications, especially when no influence on camera orientation is possible, several situations pose challenging problems. Objects and scene clutter might occlude the surface up to total occlusion (Figure 3.13). The camera might be oriented in a way that the surface is not visible at all. Though these situations are only temporal in practice, they surely lead to tracking loss if not noticed and treated. This section presents a robust filter to track the ground surface under such conditions.
Certainly, an obvious solution in these cases would be to discard the erroneous measurement and propagate the prediction until valid measurements can be gathered again. The estimated camera transformation is subject to drift in orientation as well as translation, which will propagate as error to the parameters. If the accumulated error grows over a limit, the optimization will not converge anymore when the plane comes back into view. Slight errors in plane orientation lead to quickly growing errors in the plane translation parameter d. This is visualized in Figure 3.14. If this drift remains uncorrected, the support point selection for plane fitting becomes erroneous and eventually the plane track cannot be recovered.
Interestingly, this problem is usually neglected in related systems. Sometimes the system design provides enough constraints on camera orientation and prevents the case of the surface moving out of view, as e.g. in car mounted setups. In unconstrained setups, the ground is usually assumed to be the most prominent surface and tracked by independent detections via random sampling (e.g. [55, 74]). When the ground is not visible, another prominent surface will be picked instead. Ground orientation gets lost in this moment. When elevation over ground is used for obstacle detection, this leads to total failure. In [105] we have proposed to combine two independent measurements to this end. The commonly applied plane fit in depth data, and the vertical vanishing direction measured from image evidence. In the case of a calibrated pinhole camera, this direction coincides with the normal vector of the (non-inclined) ground plane.

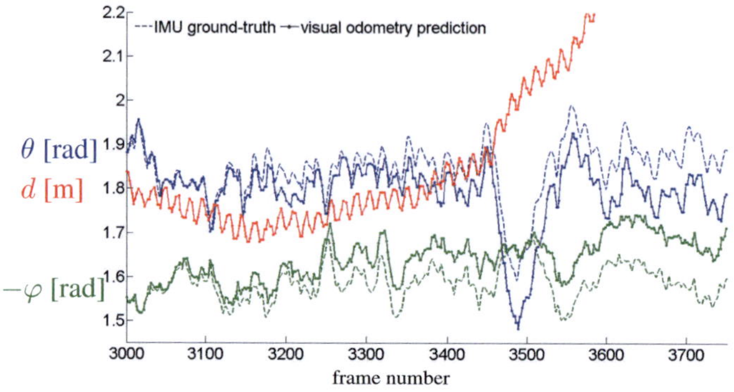

Figure 3.14: Ground plane parameters $(\mathbf{n}(\theta, \varphi), d)$ propagated by visual odometry. Without correcting drift in the attitude parameters θ and φ, the error in plane distance d grows with increasing rate. Dashed lines are the IMU ground truth, solid lines the predicted parameters. Ground truth for d is not available, the true camera height oscillates around 1.8 m.

These measurements are highly complementary: First of all, we are able to measure the vanishing direction also when the plane is not in the viewable range of the camera. Moreover, in inner-urban environments with high building facades this measurement is in principle the more accurate, the further the camera points away from the ground because more of the building structure becomes visible. Though the plane distance is not measurable from the vanishing direction alone, two of the three degrees of freedom are recoverable. On the other hand, when the camera is directed towards the floor, the depth data allows to extract an accurate plane fit which includes the plane distance.

We want to follow the principle of Section 3.1.1.3 and use both measurements to correct the plane prediction, which we obtain by propagating the parameters with visual odometry. Since we are solely interested in the current best estimate for the local plane $\pi = (\mathbf{n}(\theta, \varphi), d)$, we apply a recursive estimator to fuse the information. An extended Kalman filter fullfills our requirement here. The state of the filter are plane parameters in camera coordinates, where the plane normal vector is expressed in minimal form with spherical coordinates. We have $\mathbf{x} = (\theta, \varphi, d)^{\mathrm{T}}$. In the remainder the formulation of [85] is adopted, in which $\mathbf{x}_k^-$ denotes the estimate of $\mathbf{x}$ before incorporating the measurement of time k (predicted state), $\mathbf{x}_k^+$ the estimate after incorporating the measurement. The local plane parameters change from camera pose to pose according to the egomotion T. The egomotion can be understood as a control input, which transforms the plane parameters to the current view. We do not employ a camera motion model at this point, so the egomotion estimate constitutes the prediction step of the filter.

$$\mathbf{x}_k^- = f(\mathbf{x}_{k-1}^+, T_k) = f\left([\theta, \varphi, d]^{\mathrm{T}}, T_k\right) \tag{3.21}$$

$$= g_{sph}\left(\left(T_k^{-1}\right)^{\mathrm{T}} g_{euc}\left([\theta, \varphi, d]^{\mathrm{T}}\right)\right) \tag{3.22}$$

where $g_{sph}(\cdot)$ transforms the Euclidean plane representation into spherical coordinates and $g_{euc}(\cdot)$ vice versa:

$$\begin{pmatrix} \theta \\ \varphi \\ d \end{pmatrix} = g_{sph}\left(\begin{bmatrix} \mathbf{n} \\ d \end{bmatrix}\right) = \begin{pmatrix} \arccos(\mathbf{n}_z) \\ \operatorname{atan2}(\mathbf{n}_y, \mathbf{n}_x) \\ d \end{pmatrix} \tag{3.23}$$

$$\begin{pmatrix} \mathbf{n}_x \\ \mathbf{n}_y \\ \mathbf{n}_z \\ d \end{pmatrix} = g_{euc}\left(\begin{bmatrix} \theta \\ \varphi \\ d \end{bmatrix}\right) = \begin{pmatrix} \sin(\theta)\cos(\varphi) \\ \sin(\theta)\sin(\varphi) \\ \cos(\theta) \\ d \end{pmatrix} \tag{3.24}$$

The state transition function $f(\cdot)$ is thus nonlinear and requires the use of an extended Kalman filter with the state transition matrix approximated by the Jacobian $\boldsymbol{F} = \left.\frac{\partial f}{\partial \mathbf{x}}\right|_{\mathbf{x}_{k-1}^+, T_k}$.

We use the state prediction $\mathbf{x}_k^-$ as initialization to re-measure the vanishing direction and plane parameters. The plane representation is modified to $\alpha_n u_n + \beta_n v_n + \gamma_n + \delta(u, v) = 0$ with normalized image coordinates $u_n = \frac{u - c_u}{f}$ and $v_n = \frac{v - c_v}{f}$. This simplifies the observation model to

$$\begin{pmatrix} \alpha_n \\ \beta_n \\ \gamma_n \end{pmatrix} = -\frac{Bf}{d}\mathbf{n} = h_{uv\delta}\left(\begin{bmatrix} \theta \\ \varphi \\ d \end{bmatrix}\right) = -\frac{Bf}{d}\begin{pmatrix} \sin(\theta)\cos(\varphi) \\ \sin(\theta)\sin(\varphi) \\ \cos(\theta) \end{pmatrix} \tag{3.25}$$

with stereo baseline B and camera focal length f.

The correction step of the filter consists of two parts. The plane estimation (p. 22) provides the measurement $\mathbf{m}_{LS} = (\alpha_n, \beta_n, \gamma_n)^{\mathrm{T}}$. The vanishing point estimation (p. 31) results in a measurement $\mathbf{m}_{VP} = (\theta, \varphi)^{\mathrm{T}}$. The measurements are related to the filter state according to

$$\mathbf{m}_{LS} = h_{uv\delta}(\mathbf{x}_k^-) \tag{3.26}$$

$$\mathbf{m}_{VP} = \begin{pmatrix} 1 & 0 & 0 \\ 0 & 1 & 0 \end{pmatrix} \mathbf{x}_k^- \tag{3.27}$$

While $\mathbf{m}_{VP}$ is always applied, $\mathbf{m}_{LS}$ is considered depending on the camera pitch angle and the number of inlier points. To filter outlier measurements, both correction steps are validated with a gate around $e^2 = \mathbf{r}^{\mathrm{T}} S^{-1} \mathbf{r}$, where $\mathbf{r}$ is the update residual and S the innovation covariance [85].

3.1.3.2 Building Facades

The scene background is described by an arrangement of flat surfaces, modelled with geometric planes in Euclidean space. These planes are enforced to be orthogonally aligned with the ground plane. Thereby we take the typical gravity alignment of man made structures into account and at the same time simplify the optimization of plane parameters from three to two degrees of freedom.

The plane fitting cost function is extended by a constraint that enforces the inner product of the Euclidean plane normal vectors to be zero. Again, the optimization is kept in $uv\delta$ image space to avoid the non-linear reconstruction error. We minimize

$$\begin{gathered} \min_{\alpha,\beta} \sum_{i=1}^{N} (\alpha \cdot u_i + \beta \cdot v_i + \gamma + \delta_i)^2 \\ \text{subject to } \mathbf{n}_x(\alpha f) + \mathbf{n}_y(\beta f) + \mathbf{n}_z(\alpha c_u + \beta c_v + \gamma) = 0, \end{gathered} \tag{3.28}$$

where $\mathbf{n}$ is the ground plane normal vector. The constraint is derived from the plane transformation between image space and camera space in (3.7) (p. 23). To initialize planes vertical to the ground we modify the 3-point RANSAC method. Vertical plane hypotheses are created from two $uv\delta$ points sampled

Figure 3.15: Examples of vertical structure modelled as planar surfaces. **Right:** During initialization new hypotheses are checked for plausibility. The red hypothesis is rejected, since it would occlude the green surface in the background.

from all current non-background points and the orthogonal ground plane normal vector. Planes are tracked following the methodology of Section 3.1.1.3, but applying the constrained optimization (3.28). Other than the ground surface, vertical surfaces have distinct extents and should not be modelled as unlimited planes. For each plane we estimate the horizontal range that encloses its inlier $uv\delta$ points [49]. Together with the limiting ground plane and a fixed height over ground, a rectangular planar 3D patch is determined that is stored in the environment model. The projection of the plane patches to the image space enables to check pairs of planes for mutual occlusion. Impossible configurations can be filtered during initialization, as shown in Figure 3.15. Planes are deleted from the environment model after they could not be remeasured for a few subsequent frames.
In certain applications it might be of interest what kind of surfaces are represented by the planar patches. In [118] we developed a classifier that classifies each surface into the classes building facade, vegetation or fence. A combination of texture features (autocorrelation in horizontal and vertical direction, image gradient histograms, local binary patterns) was extracted to train a multi-class support vector machine on a dataset of 1500 labeled image patches. For the three class problem a misclassification rate as little as 1.3% could be reached. For an in-depth evaluation refer to [118].

3.1.3.3 Evaluation

The environment model should be evaluated under different aspects. An obvious quantitative measure for the performance of the scene background

estimation is the parameter accuracy and precision. This is first addressed in an empirical analysis of the variances of plane parameters. These are used to parametrize the process and measurement noise of the ground plane filter. Afterwards, the accuracy gain, that is achieved by considering vanishing point information, is quantified.
While parameter accuracy is a performance indicator, another crucial aspect is robustness. For an unconstrained system, the ability to maintain the ground reference is of particular importance, since it provides the basis for almost any succeeding perception tasks. A lost ground plane reference might lead to complete system failure. Foremost, this requires the ground plane parameter estimation to be robust against vanishing amounts of inlier points up to complete invisibility. To set challenging conditions, all datasets we use during the evaluation are captured with a head-worn camera setup (Section 4.1), which features substantial egomotion.
A last aspect that is analysed regards the completeness of the scene model. The estimated surfaces should explain all parts of the scene background, but should also not lead to an over-simplified representation that crops parts of the scene foreground.

Parameter Variance We empirically estimate the measurement variances of the least-squares parameter fit for $\alpha_n, \beta_n, \gamma_n$, the vanishing point direction $\theta_{VP}, \varphi_{VP}$ and the uncertainty in estimated camera motion in a set of experiments.

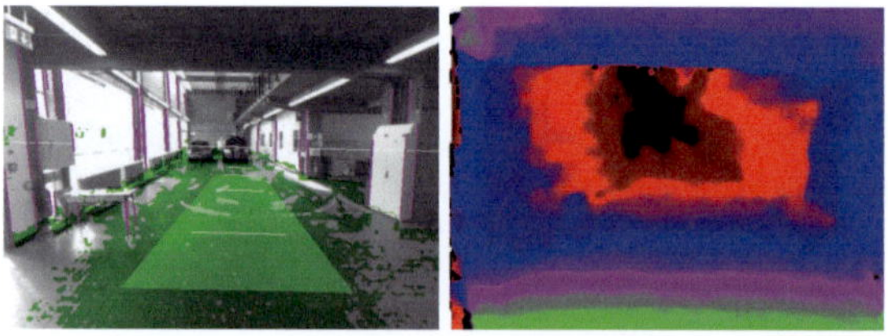

Figure 3.16: Measurement setup for parameter variance estimation. **Left:** Plane support points are colored green, edges supporting the vertical vanishing point are highlighted pink. **Right:** Corresponding disparity image.

For the $uv\delta$-plane fitting we re-estimate plane parameters 1000 times from a fixed camera position (see Figure 3.16), using the same initial solution for the support point selection. In the same manner we obtain variances for $\theta_{VP}, \varphi_{VP}$ by re-measuring the vertical vanishing point from an initial solution.
For a $uv\delta$-plane fitted in a half-resolution SGBM disparity image the parameters vary with
$\sigma_{\alpha_n} = 0.038 \quad \sigma_{\beta_n} = 0.045 \quad \sigma_{\gamma_n} = 0.016.$
The vanishing point direction varies with
$\sigma_{\theta_{VP}} = 0.0007 \quad \sigma_{\varphi_{VP}} = 0.0004.$
These values can be used as an approximation to parameterize the measurement noise in the Kalman filter. Obviously, the parameter variances are influenced by the scene itself, for instance by the size of the visible floor area, or the amount and length of visible edge segments. These factors were not considered here. The filter process noise needs to cover the uncertainty in parameter prediction, which corresponds to the uncertainty in the estimated camera motion. An approximation is discussed in Section 4.1.

Ground plane accuracy To assess the accuracy we compare the estimated plane normal to the vertical axis of the gravity-referenced IMU. For each filter update we measure the absolute deviation in degrees from the IMU ground truth for both attitude parameters $\mathbf{n}(\theta, \varphi)$. We evaluate the tracking over a sequence of 4740 frames with different disparity estimators, Table 3.1 shows the mean and standard deviation. As baseline we compare our result to the mere least-squares plane fit (LS) on the predicted plane, this corresponds to conventional ground plane fitting methods as e.g. in [81, 57, 55]. Our results after adding the vertical vanishing direction are shown in the highlighted box (LS+VP). Both attitude parameters benefit significantly from the vanishing direction, widely independent of the choice of the disparity estimator. Particularly the roll parameter φ becomes more accurate. The distributions of parameter deviations are shown in Figure 3.17, where this also becomes apparent. The effect can be explained by different facts. First, the ground plane is often occluded by cars or buildings in lateral direction, which can lead to imprecise support point selection and consequently an

	θ	φ
SGBM plane fit (LS)	$\mathbf{1.25 \pm 1.04}$	$\mathbf{2.04 \pm 1.21}$
plane fit + vanishing point (LS+VP)		
SGBM [39]	$\mathbf{0.64 \pm 0.54}$	$\mathbf{0.53 \pm 0.69}$
libElas [29]	0.50 ± 0.40	0.52 ± 0.52
libToast2 [73]	0.82 ± 0.62	0.54 ± 0.65
libToast2 w/o subpix	0.81 ± 0.65	0.53 ± 0.59

Table 3.1: Mean absolute angular deviation of ground plane normal from IMU ground truth. Errors are given in degrees $\pm$ standard deviation and evaluated for different disparity estimators. Conventional least-squares plane fitting is given as baseline.

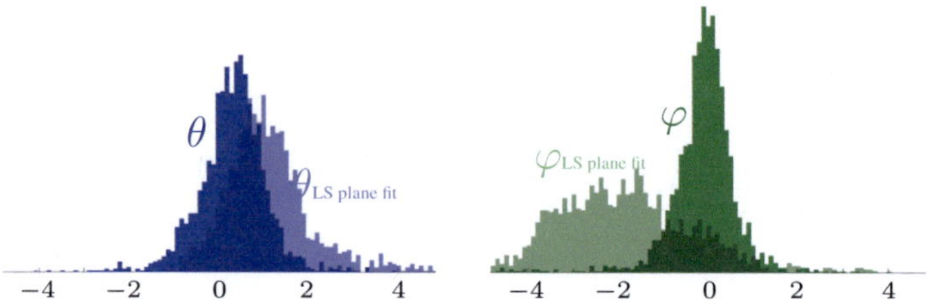

Figure 3.17: Distribution of angular error of θ and φ in degrees with and without inclusion of vanishing direction. The roll parameter φ benefits most, since objects in lateral direction often disturb the least-squares fit.

erroneous least-squares fit. Secondly, the planar surface assumption often does not hold perfectly (e.g., when walking on a slightly elevated pavement) which results in tilted measurements. See Figure 3.19 for an example. This section of the dataset lasts around 1000 frames. The absolute errors here are 1.57° for θ and 3.08° for φ with the baseline approach and decrease to 0.73° for θ and 0.41° for φ after adding the vanishing direction. Finally, the filtered parameters for the whole dataset are plotted in Figure 3.18.

Ground plane stability To test the long-term stability we ran the ground plane estimation on a dataset consisting of 45 minutes walking through inner-urban scenes, which was mostly captured on narrow side-walks between house facades and cars, but also contains some vast spaces with persons and objects frequently occluding the free view onto the ground. The lighting

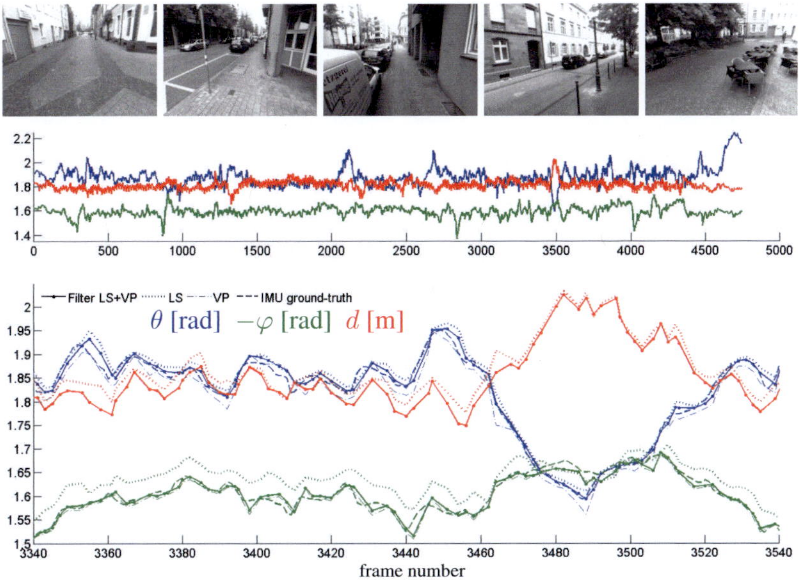

Figure 3.18: Ground plane parameters (θ, φ, d) for the evaluation dataset. Shown is the least-squares ground plane fit (LS), the vanishing direction (VP), the filter result and the IMU ground truth. The top row contains exemplary frames from the sequence.

Figure 3.19: Effect of planar surface model violation with (middle) and without (right) vanishing direction measurement. Ground plane support points are colored green and the plane is overlayed schematically. The IMU ground truth virtual horizon is drawn in red, the measured virtual horizon in green. The according error distribution for a sequence of 1000 frames around the depicted scenario is shown on the left.

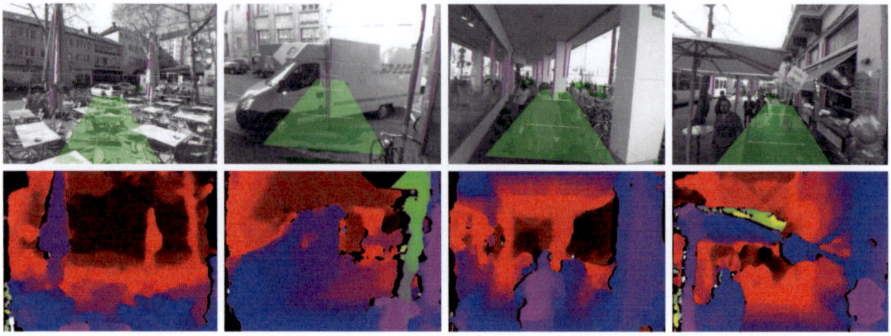

Figure 3.20: Example scenes from the city dataset. Bottom row shows the corresponding disparity measurements.

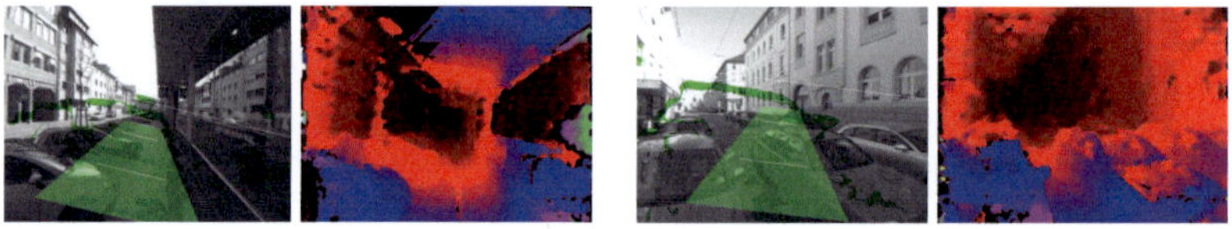

Figure 3.21: Typical failure examples without vanishing point correction.

conditions are challenging with considerably over- and underexposed image areas that lack disparity measurements. Figures 3.20, 3.21 and 3.22 contain some example shots. The method is able to keep track of the ground plane throughout the whole sequence (except for major violations of the continuous, non-inclined surface assumption, as e.g. on stairways).
The dataset we used for quantitative comparison does not contain scenes with total occlusion or situations where the ground plane is out of view. This is different here and leads to various situations in which the baseline least-squares approach looses track. In the right scene of Figure 3.21 a passing car blocks the view onto the street for a series of frames and the disparity clutter causes the plane to drift away. The vanishing direction effectively helps to keep the correct plane attitude in such cases. Figure 3.22 shows a sequence with the ground out of view due to a high camera inclination.

Figure 3.22: Sequence with ground plane not in free view due to camera inclination (third image). Only the vanishing direction is considered here to correct the plane parameters.

Without considering the vertical vanishing direction, the uncorrected drift in parameter prediction leads to tracking loss in this situation. The filter dismisses the erroneous plane fit and uses only the vanishing direction in such case.

Vertical structures The methods that are applied to estimate vertical surfaces are essentially similar to the methods used to track the ground plane. Nevertheless, they have to cope with few important differences.
The first major difference is that vertical surfaces are way more often subject to discontinuities. There is no single flat background surface as could be assumed for the ground. The background is composed of multiple surfaces with limited extent which appear and vanish as one moves through the scene. This adds the difficulties of multi-model estimation problems. It is not sufficient to track the surface parameters, additionally a mechanism is required to initialize new and remove vanished surfaces from the model.
A second difference is that background structure is much more variable in form and appearance. Even building facades are far less flat than the ground. Doors and windows are embedded, and ornaments, window frames, sunblinds or balconies stick out of the facade. Neighboured buildings often show unobvious discontinuities. Suchlike effects apply for open structures like bushes and fences. In summary, the modelling of vertical scene background as planar surfaces is of noticeably higher abstraction than it is for the ground surface in a local, limited area.
Other differences affect the measurement of plane model parameters. The distances in which background structures have to be measured are much larger. The ground stays in a camera distance of around 2 m, but building

facades face to face are often more than 10 m apart. The depth resolution of reconstructed points becomes smaller with increasing distance. Considering a camera baseline of 20 cm and a disparity estimator with a typical sub-pixel refinement of 0.125 pixels, the depth resolution in only 25 m distance already amounts to 1 m. Even if planes are measured in $uv\delta$ space, the disparity discretization limits the expectable accuracy of fitted plane models.
The estimated vertical structures are evaluated under the aspects of accuracy and completeness. While accuracy is required to apply a tracking by optimization scheme, the completeness of the background model directly influences the precision of any foreground object detector. Missed background structure will lead to the initialization of false foreground objects. A typical scenario are street canyons, where the scene is laterally delimited by high, parallel buildings. The expected environment model consists of two planes, which intersect in a small, constant angle and have a constant distance to each other. Figure 3.23 shows a few examples with the progression of intersection angle and distance over time. The distance varies in a reasonable constant range, the intersection angle varies by quite a few degrees. Both parameters stay in a range that allows to track the models by optimization over periods of several hundreds of frames. Track length is usually limited by discontinuities on the surface as e.g. the vegetation in the third image. The angle and distance distribution over all 5000 frames of this dataset is plotted underneath and reveals a similar picture.
To assess the completeness of the representation a sequence of 2200 frames in a typical urban scenario was evaluated regarding the number of missed detections and false plane models, that were initialized in scene clutter. Figure 3.24 shows some examples. In this sequence a recall of 91% is reached at a precision of 95%, speaking for a small number of false detections. The missed detections (which affect the recall) mainly occur during fast camera turns, in large distances, or when new walls enter the visible scene in a very shallow angle. Some of such typical failure examples are shown in Figure 3.25. As will be shown in Section 3.2, the detection and error rate is well sufficient in order to dismiss the scene background for local foreground object handling.

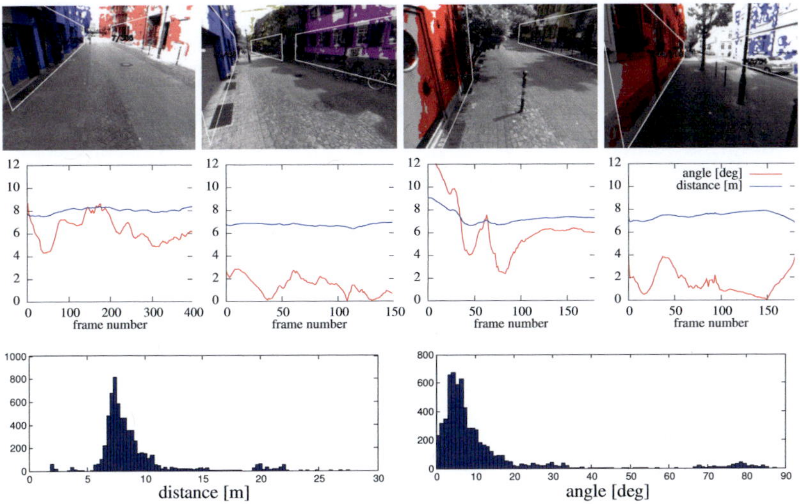

Figure 3.23: Top: Progression of intersection angle and distance of tracked planes in street canyon scenarios. **Bottom:** Distribution of intersection angle and distance over all 5000 frames in this dataset.

Figure 3.24: Examples from the evaluated inner urban sequence.

Figure 3.25: Failure examples. The first three images show examples of false detections on vegetation, the fourth image shows a missed detection on the right facade due to a very shallow viewing angle after turning around the corner.

3.1.3.4 Conclusions

The methods introduced in this section build a model of the geometric scene background. It consists of a common ground plane complemented by multiple vertical planes which represent scene delimiting structures like buildings, fences and bushes. A method was introduced to track plane model parameters through situations of total invisibility. Keeping track of the common ground also in such conditions is a fundamental ability for freely moving systems. Often, this problem is circumvented by constraining the camera orientation. We avoided this difficult restriction by extending the classical plane fitting in depth data with the complementary visual feature of vanishing points. The estimated vertical surfaces effectively cover all vertical background scene structure and allows to classify depth and image data into scene foreground and background. This will constitute the basis for obstacle detection presented in Section 3.2.
The accuracy of estimated vertical structures is principally limited, mainly by the camera baseline and image resolution, which lead to vanishing disparity gradients with increasing distance. In the evaluated camera setup (Section 4.1) this becomes apparent already in distances of 20 meters. The scenes in Figure 3.24 were previously evaluated in the context of vanishing directions (Figure 3.11 (5,6) (page 45)). The intersection angle between these planes deviates between 10° and 20° from the measured ground truth. The estimated vanishing direction model is far more accurate in suchlike scenes with large measurement distances. Hence, a logical extension of the methods presented here would consider the horizontal vanishing directions. First, in order to further constrain the model initialization with plausible surface normals, and second, to support the parameter tracking in the same manner as applied for the ground plane. Currently, there are no relations between the estimated vertical surfaces, despite sharing a common ground. A global reasoning step could connect single surfaces into a closed geometry. While impressive work with similar ideas exists for single viewpoint problems (e.g. [36]), consistent modelling over time and real-time ability are challenges that have to be overcome.

3.1.4 Stairways

The perception framework models the environment as a flat world. Considering the limited perception range this is usually a valid representation in urban scenarios. A major violation is posed by multi-level environments. Regarded level by level our model can represent such environments. Transitions between floors, typically in form of stairs, can neither be described correctly by the planar ground, nor by vertical structures, nor by foreground objects. In this setting, stairways are a special type of navigable scene background that requires individual treatment.

The detection of stairways offers valuable high-level knowledge to a variety of different intelligent systems. It is vital for systems that need to expand their range of operation to multiple floors. Systems that assist visually impaired people can provide guidance towards a stairway and provide helpful information as for instance the number of remaining steps. Not least, systems that rely on a flat world assumption need to recognise the traversal of a stairway passage to ensure proper functionality. Besides detecting the presence of a staircase also some of its properties are relevant to know. To traverse a stair the alignment, or traversal direction, the number of steps and the step height and depth have to be known. These parameters define a geometric stair model that can fully describe most regular, non-circular stairs. In this section an algorithm is detailed which detects staircases and continuously estimates these model parameters to enable a stair traversal.

Regular stairs are an assembly of treads with prominent edges which are most often aligned with the ground surface. These distinct features have been used in various approaches to detect stairs in camera images. In the early works of [80] and [64], assemblies of line features are detected with a Canny edge detector followed by Hough transform. [38] applies a Gabor filter that responds to the periodic nature of stairs. [92] and [52] trained classifiers based on Haar features to find stairs in images. Approaches that also assess the stair geometry make use of stereo vision or laser scanners. Here, stairs are measured either by detecting their tread or riser planes in the 3D data, or by reconstructing their edges. [58] proposes a curvature index to classify edges into concave and convex using a depth image. Similar is the work of [99], in which concave and convex 3D edges are extracted directly from

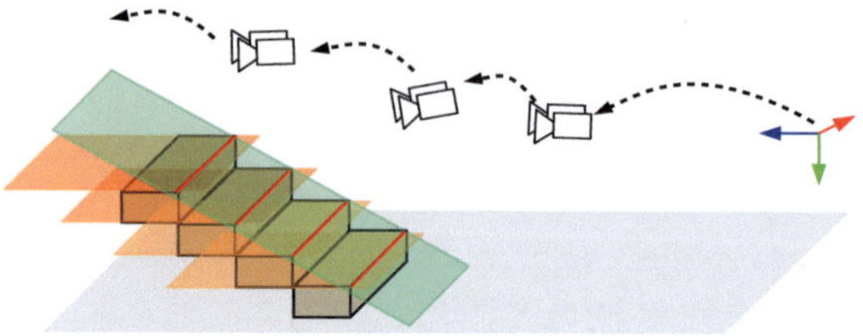

Figure 3.26: The stair is modelled with a stair plane (green) and tread planes (orange) that are evenly aligned with the ground plane.

stereo disparity data. As in [20], the edges are mapped over time. Alternative methods measure the tread surfaces. [67] compares two plane segmentation methods especially for use in sparse lidar point clouds. [72] and [69] estimate point cloud normals to segment the data into planar patches by clustering or region growing. The primitives – 3D edges, or planar patches – are then connected with some heuristics to form a stair, and if required estimate the parameters of a generative geometric model.
The focus of this work is threefold. Stairs need to be robustly detected from distance to initialize a minimal generative stair model. Instead of estimating the parameters from a single viewpoint, as focused by almost all related approaches, the stair model should be refined with the more accurate measurements that become available during stair traversal. For online applicability especially a light-weight measurement principle is required which can robustly deal with the noisy low-resolution disparity depth data. We have published early versions of the developed method in [111, 119].

3.1.4.1 Stair Model

Our staircase model consists of a plane that lies on the convex step edges (see Figure 3.26, further referred to as *stair plane*), the step height, and the number of steps. We assume the treads to be evenly aligned with the ground floor. Then, the stair inclination, or the step depth respectively, is given with the ground plane as reference.

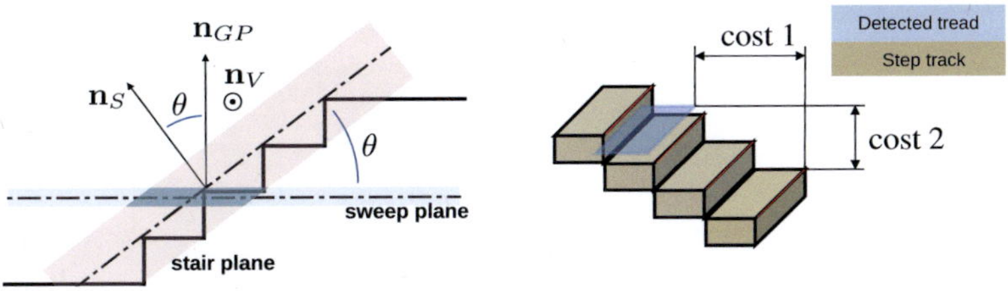

Figure 3.27: Left: Geometric relations in the stair model. The highlighted region of interest for sweep plane evaluation is given by intersecting the sweep plane and stair plane margins. **Right:** For plane association over time the horizontal as well as the vertical offset is taken into account.

Line segments (see Section 3.1.2.2), clustered according to the scene vanishing directions, are used for the initial detection of a stair. Stair planes are sampled in disparity data that corresponds to image line segments. A valid stair plane intersects the ground plane with an angle between 20° and 40° and features a required minimum number of support points. After a few successful re-detections the stair plane is tracked using continuous parameter optimization as introduced in Section 3.1.1.3. The vanishing direction of the line segments $\mathbf{n}_V$ corresponds to the intersection vector of ground plane and stair plane. Ground plane normal and stair plane normal are related by the angle axis rotation $R(\mathbf{n}_V, \theta)$, with a stair inclination angle θ (Figure 3.27). In order to traverse the stair, knowledge about step height and the number of steps is required. Related work with depth information tackles this in two different ways: Either the prominent concave and convex edges are reconstructed [58, 20, 99], or the scene is segmented into individual planar patches which are then combined to form the stair [72, 67, 69]. Particular difficulties arise when oversegmentation occurs, or when treads are split into multiple segments due to objects which occlude parts of the stair. To avoid handling these cases and spare the computational cost of full depth data segmentation, we propose a sweep plane approach to directly estimate the height of individual stair steps. We sweep the ground plane along its normal direction in discrete steps and count the support points for each position in the disparity data (Figure 3.27). The peaks of the originating profile correspond to the height over ground of each tread (Figure 3.28). This method exploits

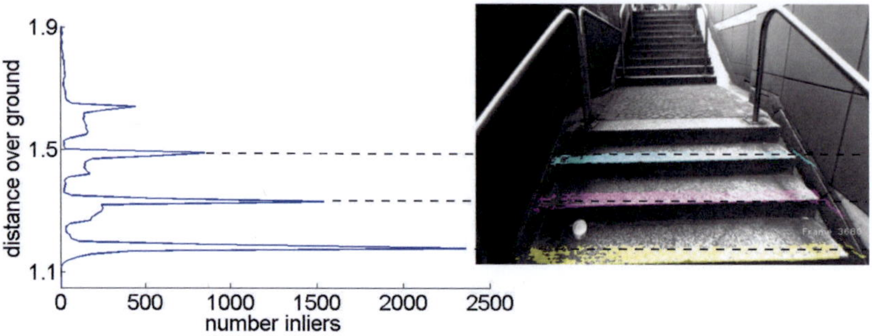

Figure 3.28: Sweep plane evaluation and corresponding image points: the left graph shows the number of points supporting the plane which is swept along the ground plane normal vector. The right image overlays the support points for the detected peaks.

the assumption that stair treads are aligned with the ground. An important prerequisite is knowledge of the correct ground plane normal, also during stair traversal when the ground plane is not visible any more. The ground plane tracking method in Section 3.1.3.1 uses the vertical vanishing direction to avoid the accumulating drift in these situations and is well suited here. In unstructured environments without clear vertical structures, the ground plane normal can alternatively be obtained from the stair plane normal and the stair inclination as mentioned earlier.

Long stairs can usually not be fully measured from a single viewpoint. Principally, the surfaces of treads located above the camera are not visible. In order to estimate the number of steps in such cases and refine the model parameters over time we track the individual treads. Sweep plane measurements are associated to the globally referenced stair model by Hungarian assignment [47]. The assignment cost is calculated as sum of the difference in height over ground and horizontal distance to the step edge (compare Figure 3.27 (right)). New stair treads are initialized from unassigned measurements.

3.1.4.2 Evaluation

The distance in which a stair is initially detected mainly depends on the lighting conditions which influence the amount of extracted edges. With the camera setup used throughout this work (Section 4.1) it succeeds from about 7 m distance. When the tracked stair plane comes into a 3 m range the sweep plane measurement is initiated and each step is tracked individually by its tread plane. In Figure 3.29 we show tracking results on an outdoor staircase. Figure 3.30 shows the step heights after tracking the stair over the sequence of 120 frames. Despite that each step is tracked individually without enforcing a constant step height, the estimated treads are equally spread.

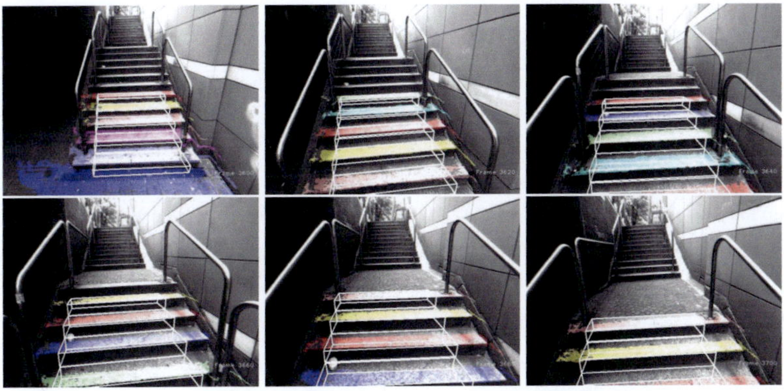

Figure 3.29: Tracking of a staircase: Visible step tracks are overlayed with different colors to indicate their identity. Cubes visualize position and size of measured steps.

Figure 3.30: Estimated treads after passing the stair in Figure 3.29 with 11 steps. Left graph shows the height of each step w.r.t. the ground (first tread). The number of detections per tread is shown on the right. Since the lower steps move out of view as the camera is moving upstairs, they are detected less often than the higher ones.

For quantitative evaluation a dataset of nine different stairs was recorded and their ground truth step height and depth manually measured. The dataset contains indoor as well as outdoor scenarios to depict varying light and stair size conditions. All stairs were recorded during ascend, three stairs additionally during descend. For each image in the sequence of a tracked stair two quantities are measured: The step height using the sweep plane method, and the stair inclination which is the angle between ground plane and stair plane. The step depth is not measured but can be derived from the inclination. The averaged value of step height and step depth are compared to ground truth in Figure 3.31 for all 12 sequences.
Over the dataset the step height is estimated with an accuracy varying between 0.1 and 1.5 cm (mean 0.7 ± 0.5 cm). The average step height in the dataset is 16.5 cm. Steps of real stairs are usually not distributed perfectly equal and easily vary in their height by a few millimetres. The estimation accuracy of the algorithm falls into the range of acquired ground truth accuracy.
The step depth estimation shows an accuracy reaching from 0.1 cm up to 4 cm (mean 1.6 ± 1.3 cm), hence being less accurate than step height estimation. The inaccuracy results from a stair plane inclination error of $1.18 \pm 0.9°$ on average. It can be explained by the influence of disparity clutter in the stair vicinity and lacking or spurious edge measurements due to low light conditions or cast shadows.

	Step Height (cm)	Step Depth (cm)
Proposed method	0.71 ± 0.46 (4.4%)	1.56 ± 1.30 (5.4%)
Edges, Kinect [20]	1.7 ± 1.4 (8.9%)	1.2 ± 1.6 (4.2%)
Edges, Stereo [99]	0.12 ± 0.66 (0.8%)	0.24 ± 1.14 (0.85%)
Planes, SLG, Lidar[67]	0.42 ± 0.31 (6.0%)	1.17 ± 0.67 (6.5%)
Planes, TPRS, Lidar [67]	0.68 ± 0.54 (9.7%)	0.90 ± 0.61 (5.0%)
Planes, RG, RGB-D [69]	1.44 ± 0.59 (8.5%)	0.61 ± 1.89 (2.0%)

Table 3.2: Mean absolute error $\pm$ standard deviation, as well as error percentage of stair slope and step height: The comparison is performed between our method, and reported accuracies of the 3D edge-based method in [20] and [99], plane-based scan-line grouping (SLG) and two-point random sampling (TPRS) from [67], and planar region-growing (RG) in [69].

A comparison of estimation errors between our method and the reported results of [67], [20], [69] and [99] are presented in Table 3.2. The results are difficult to compare. There is no commonly used benchmark dataset and the approaches differ in the type of depth sensor. Furthermore, the height of the viewpoint differs which has a big influence on how much and which parts of the stair are visible. Despite working on low resolution disparity data, the accuracy and precision of the proposed method is well on a level with related approaches based on way more accurate depth data from laser scanners, RGB-D cameras, or higher resolution disparity data. For many of the evaluated stairs our results are more accurate. Little can be said about the repeatability and robustness of related approaches, since commonly a maximum of 3 different stairs is evaluated. These are usually of the same kind and primarily leading upstairs. Principally, the accuracy can be expected higher here, due to smaller measurement distances and the full visibility of treads.

After all, the step dimension estimation is only one part of the problem. The primary focus here lies on traversing the stair. Besides a stable initial detection of the staircase itself this requires to keep track of the stair plane and the location of steps, also those which only come into view at a later point in time.

During walking in urban scenarios, the initial detector for staircases produces around 1 erroneous detections per 500 images. For these miss-detections usually no valid step measurements can be obtained, whereupon the stair track is interrupted quickly again within a few frames. Embedded into the whole perception framework, surfaces classified as vertical structure are removed from depth data prior to detecting stairs, which lets the error rate drop to literally no fail detections.

Other than active robotic platforms, our passive perception framework cannot influence camera orientation and movement. It is not possible to direct the attention from step to step. By contrast, given the high viewpoint of our experimental setup the immediate next two steps are usually not visible. Approaches that require the ground plane as visible reference suffer from orientation loss as soon the stair is entered. Stair traversal under such conditions is only treated here and in our related work of [99].

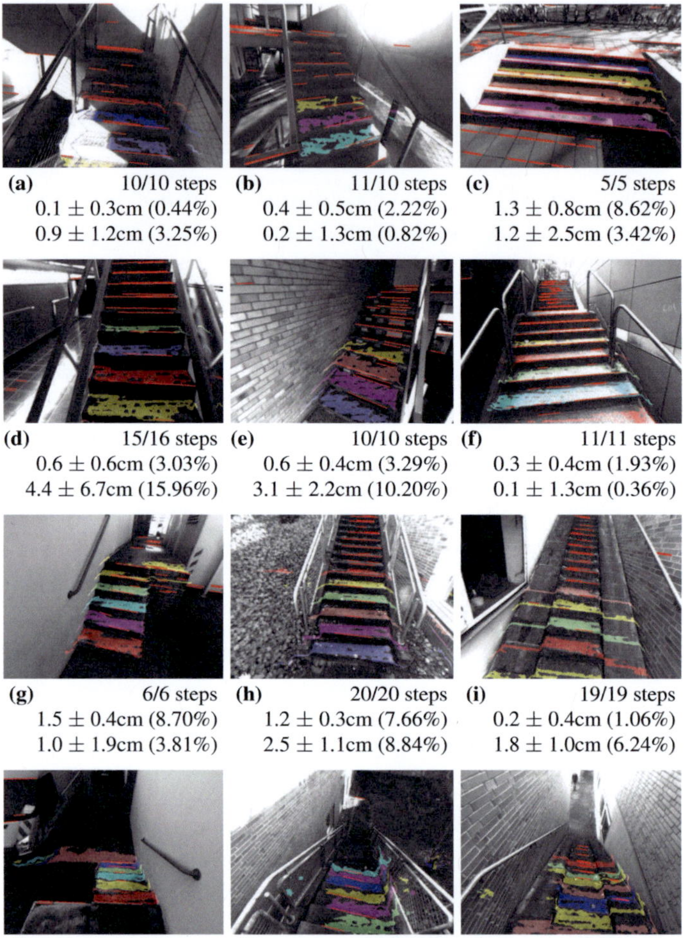

	(a) 10/10 steps	(b) 11/10 steps	(c) 5/5 steps
Height	0.1 ± 0.3cm (0.44%)	0.4 ± 0.5cm (2.22%)	1.3 ± 0.8cm (8.62%)
Depth	0.9 ± 1.2cm (3.25%)	0.2 ± 1.3cm (0.82%)	1.2 ± 2.5cm (3.42%)

	(d) 15/16 steps	(e) 10/10 steps	(f) 11/11 steps
Height	0.6 ± 0.6cm (3.03%)	0.6 ± 0.4cm (3.29%)	0.3 ± 0.4cm (1.93%)
Depth	4.4 ± 6.7cm (15.96%)	3.1 ± 2.2cm (10.20%)	0.1 ± 1.3cm (0.36%)

	(g) 6/6 steps	(h) 20/20 steps	(i) 19/19 steps
Height	1.5 ± 0.4cm (8.70%)	1.2 ± 0.3cm (7.66%)	0.2 ± 0.4cm (1.06%)
Depth	1.0 ± 1.9cm (3.81%)	2.5 ± 1.1cm (8.84%)	1.8 ± 1.0cm (6.24%)

Down-stairs

	(j) 6/6 steps	(k) 7/20 steps	(l) 13/19 steps
Height	0.8 ± 1.5cm (4.64%)	0.3 ± 2.9cm (2.00%)	1.2 ± 1.9cm (7.31%)
Depth	0.8 ± 3.3cm (2.88%)	0.1 ± 6.0cm (0.35%)	1.6 ± 4.2cm (5.44%)

Figure 3.31: The 12 stair sequences used in the evaluation. Errors of step height and depth are given as mean absolute error ± standard deviation, and the percentage of the error compared to the measured ground truth. Additionally, the number of estimated steps is given. The dataset contains closed as well as open stairs (e), and ascending (a-i) as well as descending (j-l) sequences. The stairs in (j-l) are the same as the stairs in (g-i), but are captured while moving downstairs.

Tracking the stair model succeeds in all ascending stairs contained in our dataset. Tracking the single steps fails in few cases. In b) and d) errors in the data association lead to a wrong overall step count. This is usually seen in cases where the camera is temporarily pointing away from the stair. During stair descend the method does not reach the same stability. In k) the stair plane gets lost which stops the algorithm. In l) many step measurements are wrongly associated resulting in 6 missing steps after traversal. The performance gap can mainly be explained by the larger measurement distances compared to an ascending stairway. Additionally, all step treads are partly occluded by the previous steps and hence smaller than in the ascending traversal.
The computation time of our approach during stair traversal amounts to 35-40 ms after disparity estimation and allows to track the stair model with around 12 fps. Along with [20, 69] and the scan-line grouping of [67] our approach is one of the few methods lightweight enough for online applicability on wearable hardware.

3.1.4.3 Conclusion

We presented a method to measure and track stairways using a binocular camera. Based on the assumption of evenly aligned stair treads, an efficient sweep plane method was proposed to measure the step height. Steps are tracked over time in order to refine the minimal geometric stair model during traversal. The measurement principle can be applied to ascending as well as descending stairways, and works for closed as well as open stairs that consist only of treads. The evaluation shows robustness on a wide range of diverse stairways, while achieving measurement accuracies equal to methods designed for more accurate depth sensors. Therewith, the method can be an enabler for different applications in the domain of mobile robotics or wearable perception platforms.

3.1.5 Correcting Odometry Drift

During environment modelling we treat geometry estimation and egomotion estimation as independent problems. This leads to an environment model,

which is consistent in a local sense, but is directly afflicted with the odometric drift when regarded globally. As argued before, this does usually not imply functional restrictions for tasks that concern the immediate local environment. Clearly however, model consistency becomes the more important, the wider the range of interest. Based on our previous work [104], we will show how the scene background information can be fed back into the egopose estimation to extend local consistency to the whole surrounding scene.
Errors in egopose estimation are caused by long term systematic drift, or also by single erroneous odometry readings. Consider the scenario depicted in top view in Figure 3.32. While travelling into direction A, the yaw drift causes the position to slowly deviate from the true travelled path. The platform then performs a hard turn into a side road. The error that happens during the short period of turning is propagated and cannot be recovered. A compass reference measurement would be sufficient to correct both kinds of errors in the fashion of an inertial filter. On a vision only platform we can mimic the function of a compass by measuring one scene vanishing direction, say A. Such idea has been investigated, e.g. in [44, 75], to counteract the systematic drift. Once the platform turns into direction B, direction A becomes immeasurable and the scene reference is lost. To correct the turn error, the reference needs to be switched to direction B. Yet, to guarantee scene consistency, the relation between A and B has to be known. In many constrained environments this is the case. The interior of buildings for instance most often conforms with the Manhattan assumption where all

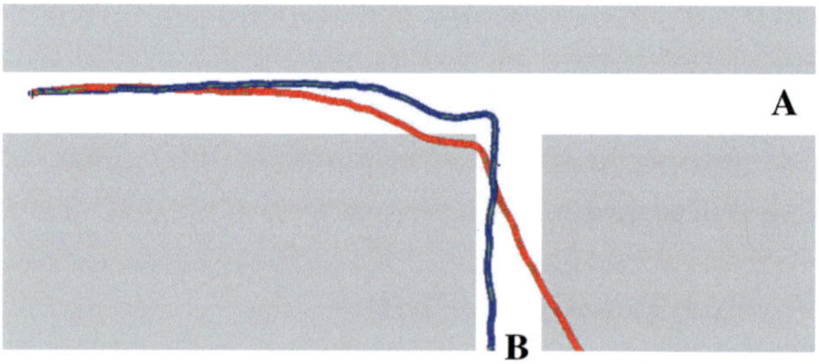

Figure 3.32: Top view of an estimated path. Systematic drift as well as short term odometry errors can cause large position errors (red). Both types can be corrected (blue) by considering the scene vanishing directions A and B.

directions are orthogonally aligned. Estimating these relations between directions in unknown, changing environments was focus of Section 3.1.2.

3.1.5.1 Visual Orientation Filter

We are interested in the attitude (orientation) and position of the platform with respect to a scene fixed coordinate system W. In discrete timesteps k we obtain odometric measurements T_k consisting of rotation R_k and translation $\mathbf{t}_k$. They accumulate to the overall transformation

$$T_{W_k} = \begin{pmatrix} R_{W_k} & \mathbf{t}_{W_k} \\ \mathbf{0}^T & 1 \end{pmatrix} = T_{W_{k-1}} T_k^{-1} = T_{W_{k-1}} \begin{pmatrix} R_k^{\mathrm{T}} & -R_k^{\mathrm{T}} \mathbf{t}_k \\ \mathbf{0}^T & 1 \end{pmatrix} \tag{3.29}$$

as earlier depicted in Figure 2.3 (p. 13). This incremental update step can be decomposed into attitude and position as

$$R_{W_k} = R_{W_{k-1}} R_k^{\mathrm{T}} \tag{3.30}$$

$$\begin{aligned} \mathbf{t}_{W_k} &= \mathbf{t}_{W_{k-1}} + (-R_{W_{k-1}} R_k^{\mathrm{T}} \mathbf{t}_k) \\ &= \mathbf{t}_{W_{k-1}} - R_{W_k} \mathbf{t}_k. \end{aligned} \tag{3.31}$$

Figure 3.33: Three vanishing directions typically found in indoor scenarios. A measurement of these directions provides a scene fixed reference that we apply to counteract the drift in odometric pose estimation. Background image: Google Earth, © 2009 GeoBasis-DE/BKG.

Notice, that estimation errors in the odometry rotation R_k are propagated into the position update. This dependency causes the largest part of accumulated position drift. Even if the velocity of the platform can be perfectly estimated, incrementing the movement into an error-prone direction leads to large position errors. By implication, correcting the rotation error with a referenced direction measurement will strongly improve the estimated position.
From two vanishing directions we could directly determine the camera orientation with respect to the scene R_{W_k}, free of incremental drift. In the usual case however, the frequency of odometry measurements will be higher than that of vanishing directions. Therefore, the odometry measurements R_k are used as (non-referenced) attitude prediction. The vanishing directions are then applied as correction to counteract the small errors. We use the Kalman filter framework to implement this as an recursive state estimator. In principle it is not possible to correct for translation drift in $\mathbf{t}_k$ with referenced direction measurements alone. Therefore, the translation update is treated independently without any drawbacks.
The global attitude R_{W_k} constitutes the filter state and is expressed as an orientation quaterion $\mathbf{q}$. The odometry orientation component R_k is received as quaternion $\mathbf{v}$. Equivalent to (3.30), the filter prediction step is given by

$$\mathbf{q}_k^- = \mathbf{q}_{k-1}\mathbf{v}^* \tag{3.32}$$

where $\mathbf{v}^*$ denotes the conjugate of $\mathbf{v}$. We can write this in matrix form as

$$\mathbf{q}_k^- = A\mathbf{q}_{k-1} = \begin{pmatrix} \mathbf{v}_0 & \mathbf{v}_1 & \mathbf{v}_2 & \mathbf{v}_3 \\ -\mathbf{v}_1 & \mathbf{v}_0 & -\mathbf{v}_3 & \mathbf{v}_2 \\ -\mathbf{v}_2 & \mathbf{v}_3 & \mathbf{v}_0 & -\mathbf{v}_1 \\ -\mathbf{v}_3 & -\mathbf{v}_2 & \mathbf{v}_1 & \mathbf{v}_0 \end{pmatrix} \cdot \begin{pmatrix} \mathbf{q}_0 \\ \mathbf{q}_1 \\ \mathbf{q}_2 \\ \mathbf{q}_3 \end{pmatrix}_{k-1} \tag{3.33}$$

and propagate the state covariance Σ according to $\Sigma_k = A\Sigma_{k-1}A^{\mathrm{T}} + \Sigma_Q$. The process noise Σ_Q is chosen large enough to accommodate the visual odometry uncertainty (an estimate is given in Section 4.1).

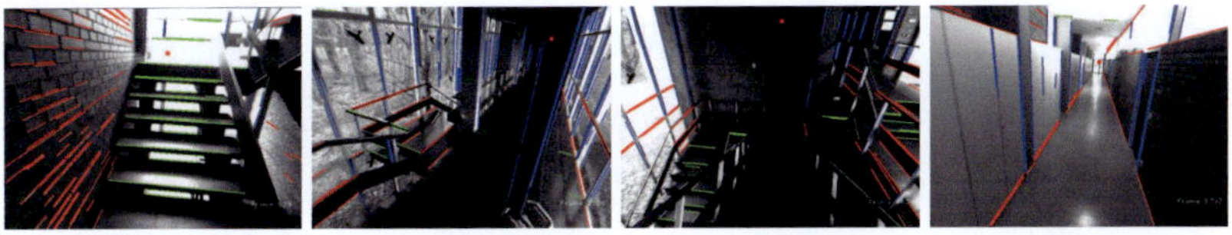

(a) (b) (c) (d)

Figure 3.34: Example shots from the image sequence. Positions are marked in Figure 3.35.

The i scene vanishing directions are fixed in W with spherical coordinates $(\theta_{W_i}, \varphi_{W_i})$. The measurement model for a direction (θ_W, φ_W) from current attitude $\mathbf{q}_k^-$ reads as

$$h(\mathbf{q}_\mathbf{k}^-, \theta_W, \varphi_W) = \begin{pmatrix} \theta \\ \varphi \end{pmatrix} = g_{sph}\left(R(\mathbf{q}_k^-) \begin{pmatrix} \sin(\theta_W)\cos(\varphi_W) \\ \sin(\theta_W)\sin(\varphi_W) \\ \cos(\theta_W) \end{pmatrix} \right) \tag{3.34}$$

where $R(\mathbf{q})$ is the left-handed rotation matrix equivalent to the rotation quaternion $\mathbf{q}$ and

$$\begin{pmatrix} \theta \\ \varphi \end{pmatrix} = g_{sph}(\mathbf{n}) = \begin{pmatrix} \arccos(\mathbf{n}_z) \\ \operatorname{atan2}(\mathbf{n}_y, \mathbf{n}_x) \end{pmatrix} \tag{3.35}$$

is the transformation between Euclidean and spherical coordinates.
For each vanishing direction we apply one correction step with the local measurements (θ_j, φ_j). The Kalman update with linearised measurement model $\boldsymbol{H} = \left.\frac{\partial h}{\partial \mathbf{q}}\right|_{\mathbf{q}_k^-}$ yields the updated attitude estimate $\mathbf{q}_k^+$.
To complete the process, the odometry translation measurement $\mathbf{t}_k$ is transformed into the local frame and incremented following (3.31) as

$$(0, \mathbf{d}) = \mathbf{q}_k\,(0, -\mathbf{t}_k)\,\mathbf{q}_k^* \tag{3.36}$$

$$\mathbf{t}_{G_k} = \mathbf{t}_{G_{k-1}} + \mathbf{d} \tag{3.37}$$

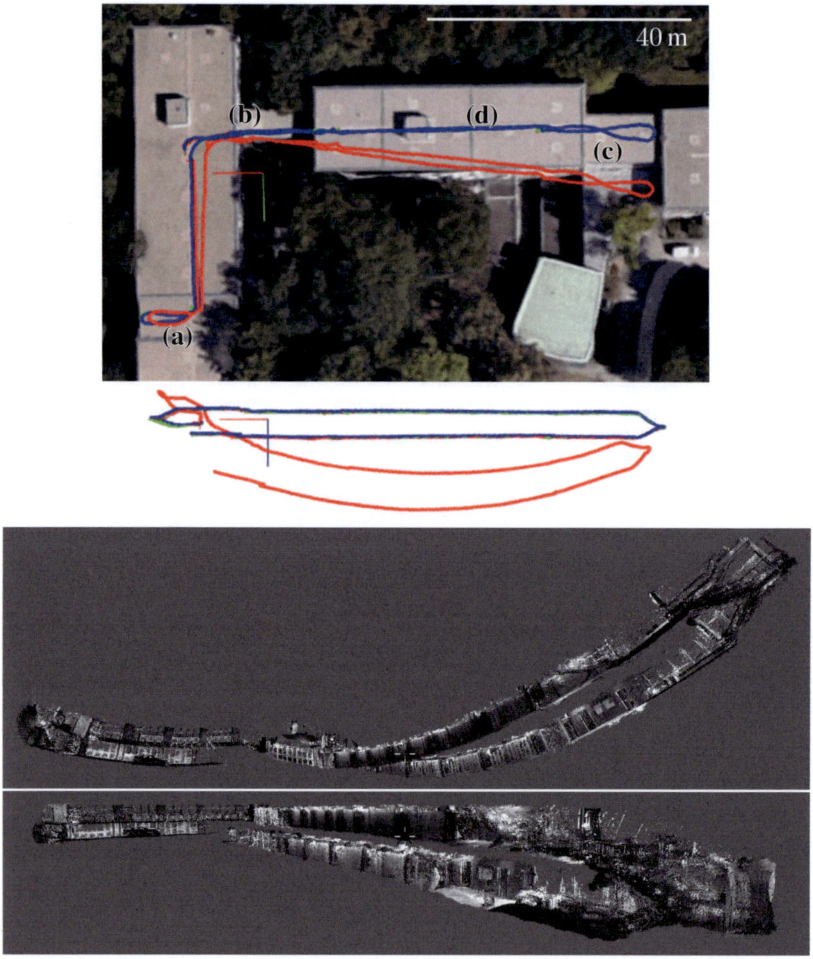

Figure 3.35: Top: Comparison of filtered trajectory in blue and plain visual odometry in red. The plotted coordinate system corresponds to the global vanishing directions. A slight drift can be seen in the upper top-view, the side-view below reveals a large drift in vertical direction. Background image: Google Earth, © 2009 GeoBasis-DE/BKG. **Bottom:** A mapping system could produce an almost consistent reconstruction of the building (before and after drift correction).

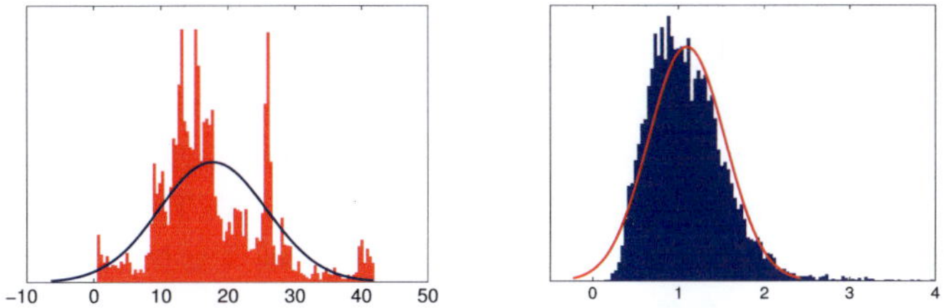

Figure 3.36: Distribution of attitude error in degrees without (left) and with (right) vanishing direction reference compared to IMU ground truth. Mind the differently scaled x-axis of the plots.

3.1.5.2 Evaluation

The introduced method is general in the sense that it can be applied on top of any odometric sensor as e.g. wheel encoders in robotic applications. In our case these measurements are derived from the camera itself by visual odometry. This is no restriction but rather the most unconstrained case, as position and orientation have to be estimated in all 6 degrees of freedom.

Experiments indoor as well as outdoor demonstrate the gain in estimation accuracy. The first scenario consists of a 240 m loop through a building including two staircases connecting the two floors as well as glass doors in the corridors that had to be opened during the passage. Opening doors is a challenging situation for visual odometry, since large parts of the observed scene are moving and violating the static scene assumption.

The achievable accuracy is limited by the accuracy of the estimated vanishing directions. The measured vanishing directions for this scenario were evaluated in Section 3.1.2.5. Figure 3.10b (p. 42) shows the measurements transformed into the IMU reference frame. The variances of the measured parameters θ and φ can be determined from this mapping ($\sigma_{\theta,\varphi} = 10^{-4}$) and are used to parameterize the measurement noise of the filter correction step. The estimated path is plotted in Figure 3.35. The horizontal drift of uncorrected visual odometry (red path) becomes obvious on the long straight corridors (top), the vertical drift appears even stronger in this sequence (bottom). The filter is able to correct this drift (blue path) by using the three orthogonal vanishing directions (corresponding to the axis of the plotted

coordinate system) as environment fixed reference. The overall position error after closing the loop decreases from 3.6% to 1.2% using the filtered estimate. Compared to IMU ground truth, the uncorrected incremental visual odometry deviates by up to 40° from the true attitude, after correction this reduces to a mean deviation of around 1° (Figure 3.36). This corresponds to the noise found in the vanishing direction estimation. The position increment benefits accordingly. A mapping system could produce an almost consistent reconstruction of the building, as is shown in the lower part of Figure 3.35. Note, that it would not suffice to detect the loop-closure in the end to correct for the drift within the loop. This correction requires a scene wide measurement. The remaining small drift can be tied to accumulating errors in the odometric translation component $\mathbf{t}_k$, which can principally not be corrected with direction references only.
The second scenario in Figure 3.37 was recorded outdoor on an approx. 400 m loop. Other than in the indoor scenario, only the vertical vanishing direction was measured here as reference to correct the odometry. Even though no horizontal reference was used, the effect is strong. With one

Figure 3.37: Comparison of filtered trajectory (blue) and plain visual odometry (red) in an outdoor scenario. Only the vertical vanishing direction was used to correct the odometry. Background image: Google Earth, © 2009 GeoBasis-DE/BKG.

direction reference we can correct two degrees of freedom (here pitch and roll). The odometry drift afflicts mainly the pitch angle in this scenario and leads to a maximal vertical deviation of up to around 40 m.
The idea of using vanishing directions as a visual compass has been implemented in different forms with differing sensor setups. In [53] a solution for measurements originating from laser scanners is presented. Implementations based on cameras are found in [43, 75]. Also camera-based are [44, 12], but integrated into visual inertial platforms which additionally use an inertial measurement unit.
The comparable approaches of [44] and [75] solely make use of the frontal vanishing direction, which they reinitialize after a turn. While they can mitigate the orientation drift in straight passages, it is impossible to correctly estimate the direction of the turn. None of the approaches makes use of the fact, that vanishing directions are fixed with respect to each other. As shown in Section 3.1.2 this facilitates the measurement itself, but also provides valuable knowledge about the scene which we exploited here.

3.2 Generic Multi-Object Detection and Tracking

Once the geometric scene background is known, the largest part of the scene can already be explained. What remains unexplained is the scene foreground, which is composed of individual objects placed within the scene. These may be of interest as goals of navigation, or as obstacles to be avoided during path planning. Modelling these objects is subject of this section.
Object handling in context of a mobile, intelligent system should be considered as a continuous problem. We are not only interested in reliably detecting the presence of an object, but also in its movement in order to react properly to dynamic objects. To this end, objects need to be recognized between sensor readings and tracked over time. More generally, the task is to continuously estimate the state of the object, which over time forms an object track. The state characterizes the object and may contain the position in the environment model, its velocity and direction of movement. Each camera image contains a measurement to refine the state estimate.
In most object tracking systems a "small object" assumption is made. Objects are assumed to be points in the measurement space, that can be modelled

without spatial extent and that generate at most one measurement per sensor readout. In the multi-object case objects are treated independently. Each sensor readout contains measurements that are generated by the objects, albeit the association is unknown. If exactly one measurement per object exists, the optimal global solution to this assignment problem is given by the Hungarian method [47]. In practice though, measurements are ambiguous. Measurements can be missing, or be falsely generated by noise. Additionally, the number of objects is unknown – objects can appear and disappear from the scene. This renders multi-object tracking a challenging problem.

Camera-based object tracking systems are usually provided with measurements from an object specific detector. Such detector provides point like measurements of the object, usually of its position and size in the image space. Associating detections over time leads to the popular tracking-by-detection scheme. It is applicable when object types of interest are limited and known in advance, which is clearly not fulfilled in our scenario. When domain and object specific knowledge does not exist, the measurements need to be provided by a partitioning of sensor data into object signal and background noise. Moving objects are often separated by evaluating the egomotion compensated scene flow (e.g. [54]), which, however, vanishes for static objects. In our case a partitioning is partly given by the scene background model. Missing is a partitioning of the scene foreground into distinct objects.

Since the characteristic appearance of objects is extremely variable and here also unknown, it is hard to exploit appearance as a segmentation feature. More general are spatial considerations, as for instance the fact that objects usually appear separated from each other. Spatial information is provided by binocular depth data, however, the measurements are of low-resolution, noisy and corrupted for any reflective or overexposed part of the scene. Objects can be arbitrarily sized, they can be occluded by other objects, or only visible partly when entering and leaving the sensing range. Under these conditions there exists no solution to reliably segment depth data into object hypotheses. Each object will typically generate many measurements, caused by the fact that real objects are no point sources in sensor space but have an unknown shape and spatial extent. In order to track such objects, their shape cannot be neglected. It is an important clue during measurement association and enables to reason about occlusion. In each timestep, typically only some parts of the object shape are measurable due to self and inter-object occlusion. A complete shape model of the object arises only after accumulating measurements over multiple timesteps, while simultaneously estimating the unknown object motion. As opposed to tracking of small point like features, the problem is known as "extended object tracking" [35].

3.2.1 Object Model

Each object O is represented by its shape $\mathbf{O}$ and its kinematic state $\mathbf{x}$. The shape is an unstructured set of 3D points $\mathbf{O} = \{x_1..x_n, y_1..y_n, z_1..z_n\}^W$ in the global reference frame W. Different forms of shape models $\mathbf{B}$ can be derived from $\mathbf{O}$ to determine the spatial object extensions. The kinematic state $\mathbf{x} = (\mathbf{P}, \dot{\mathbf{P}})^W = (x, y, z, \dot{x}, \dot{y}, \dot{z})^W$ comprises the object's 3D position and a constant velocity component in W. The state uncertainty $\Sigma_{\mathbf{x}}$ is expressed by covariances $\Sigma_{\mathbf{P}}^W$ and $\Sigma_{\dot{\mathbf{P}}}^W$. The goal is to continuously estimate the set of objects $\{O \in \mathbb{O} \mid O = (\mathbf{x}, \Sigma_{\mathbf{x}}, \mathbf{O}, \mathbf{B})\}$ with unknown and changing cardinality using measurements derived from the disparity data. Over time, each object forms an object track.

3.2.2 Measurement Generation

Measurements are groups of $uv\delta$ points $S = (u_1..u_n, v_1..v_n, \delta_1..\delta_n)$, which result from a segmentation of the disparity data. During segmentation three cases may occur. (a) A single segment represents the whole object. (b) Multiple segments represent the object (oversegmentation). (c) A segment represents multiple objects (undersegmentation). Case (a) cannot be guaranteed, since generic semantic segmentation can be considered an unsolved problem. Undersegmentation (c) causes various difficulties. It leads to initialized objects, which span over multiple actually independent instances. An explicit method is required to recognize such unintended merges and split the objects accordingly. Handling potential undersegmentation during state estimation requires the explicit option to associate multiple object tracks to the same measurement. Oversegmentation (b) on the other hand may lead

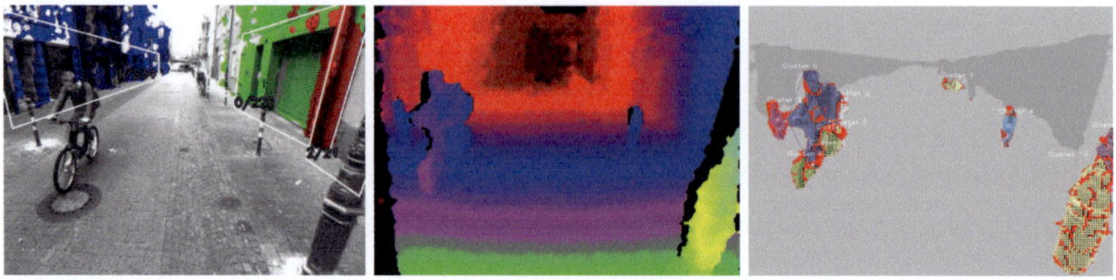

Figure 3.38: After discarding the scene background (left) from the disparity map (middle), remaining points are clustered (right). Strong disparity gradients (red) form barriers, which prevent undersegmentation.

to multiple spurious tracks initialized for the same object. In general this situation is easier to handle than case (c) and leads to less severe errors or tracking loss. Simple temporal filtering alone can mitigate many of these errors. In practice, over- or undersegmentation cannot be avoided reliably. However, by tweaking the scene segmentation towards oversegmentation, the problems of undersegmentation can be avoided and neglected. Then, each segment is known to be generated by either exactly one object, or background clutter.

A simple and fast oversegmentation is computed by single agglomerative clustering. As distance function between two neighboured pixels their disparity difference is evaluated. To avoid undersegmentation over spurious object contours, large disparity gradients are removed in advance with a gradient filter. As a result, the data is clustered into a varying number of $uv\delta$ point segments $S_i \in \mathbb{S}$, see Figure 3.38 for an example.

3.2.3 Measurement Partitioning

The fact that each object will generate multiple incomplete measurements through an unknown sensor model (the *sensor* being the segmentation algorithm) makes a direct state estimation from the measurements infeasible. An intermediate step is required which groups measurements into sets, or detections, that reflect whole objects. Essentially, each segment could have been generated by each existing object track, by a new object, or by noise.

An optimal multi-object state estimator would consider all possible sets of measurements and associations to existing object tracks. However, even for small amounts of segments, the number of possible sets leads to infeasible computational problems (more than 10^5 possible sets for as little as 10 measurements). To keep the problem tractable, meaningful set approximations are required. It is likely that segments located close to each other were generated by the same object. Furthermore, it is likely that segments aggregate close to the existing objects. This motivates us to cluster the segments based on their distance to existing objects.

The probability that a segment S was generated by an object O is approximated by evaluating the Mahalanobis distance D_M between the reconstructed segment and the object position. For normally distributed measurements the squared Mahalanobis distance D_M^2 is χ_k^2 distributed (with the degrees of freedom k equal to the dimensionality of the measurement, i.e. $k = 3$). The probability that a measurement originates from an object is then given by the cumulative distribution. A validation gate $V(\gamma) = D_M \leqslant \gamma$ can be set up around the objects by defining the required probability $P(S \in V(\gamma))$ [18]. Certainly, in the present setting this can only be seen as an approximation, since the assumption of normally distributed measurements will not completely be true. It is presumably violated by the segmentation process, which yields varying numbers of differently sized segments, and by the segment reconstruction process, which introduces a slight distortion to the normal distribution even if measurements were normally distributed in $uv\delta$ image space. Nevertheless, the Mahalanobis distance itself is an efficient measure to compare measurements under uncertainties in a consistent way. This is important especially during the association of reconstructed camera measurements, which show measurement uncertainties growing quadratically with increasing camera distance, as was previously discussed in Section 2.1.

We calculate D_M between object and segment positions in local Euclidean space (see Figure 3.39). Object tracks are maintained in global Euclidean space W with position $\mathbf{P}^W$ and position uncertainty $\Sigma_{\mathbf{P}}^W$, while segments are measured in local $uv\delta$ image space. Thus, both need to be transformed.

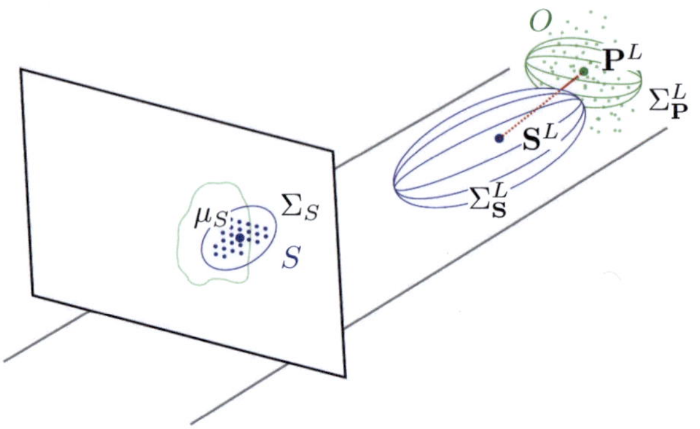

Figure 3.39: Distance (red) between an object O, positioned at $\mathbf{P}^L$ with covariance $\Sigma_{\mathbf{P}}^L$, and a $uv\delta$ segment S in image space. $\mathbf{S}^L$ and $\Sigma_{\mathbf{S}}^L$ are reconstructed from the segment centroid μ_S, considering its covariance Σ_S.

The local position of an object and the associated covariance are obtained as

$$\mathbf{P}^L = T_W^{-1}\mathbf{P}^W \tag{3.38}$$

$$\Sigma_{\mathbf{P}}^L = T_R \Sigma_{\mathbf{P}}^W T_R^{\mathrm{T}} \tag{3.39}$$

with $T_W = \begin{pmatrix} T_R & T_{\mathbf{t}} \\ \mathbf{0}^{\mathrm{T}} & 1 \end{pmatrix}$ being the current egopose estimate.

The position and uncertainty of a segment S is reconstructed from the segment centroid $\mu_S = (\bar{u}, \bar{v}, \bar{\delta})$ by

$$\mathbf{S}^L = F(\mu_S) \tag{3.40}$$

$$\Sigma_{\mathbf{S}}^L = J_F(\mu_S) \cdot \Sigma_S \cdot J_F(\mu_S)^T. \tag{3.41}$$

Σ_S reflects the position uncertainty in image space and is given as the covariance of $uv\delta$ points $\Sigma_S = \mathrm{Cov}(S)$ (Figure 3.39). $F(\cdot)$ and $J_F(\cdot)$ are the inverse camera projection function and its Jacobian (Section 2.1). This way the uncertain location of the segment centroid as well as the uncertain disparity measurement are considered.

The Mahalanobis distance is then calculated as

$$D_M(\mathbf{P}^L, \mathbf{S}^L) = \sqrt{(\mathbf{P}^L - \mathbf{S}^L)^{\mathrm{T}}(\Sigma^L_{\mathbf{P}} + \Sigma^L_{\mathbf{S}})^{-1}(\mathbf{P}^L - \mathbf{S}^L)}. \tag{3.42}$$

The segment measurements are clustered into sets G by assigning them to the most likely objects according to D_M, considering the gating volumes V as defined above. Besides building useful measurement sets, also the data association between sets and objects is solved by this step. For each measurement set G the centroid $\mu_G = (\bar{u}, \bar{v}, \bar{\delta})$ and its covariance Σ_G is determined. The sets form detections of the objects and are used in the following to update their states.

3.2.4 State and Shape Estimation

The object states are tracked in the global reference frame with independent Kalman filters using either a constant position or constant velocity model. Objects enter the detection range in around 15-20 m distance and new tracks have to be initialized. In these distances the measurement noise is larger than the apparent motion of slowly moving objects like pedestrians. Their mere distinction from static objects is difficult. In consequence, the established filter velocity components are very unreliable here. Moreover, for static objects, the constant velocity model is overparameterized and deteriorates the position estimate. Therefore, a constant position filter model is initially applied upon track initialization. Two conditions can switch the filter to a constant velocity model. Either a high percentage of motion features (see Section 2.2) within the object hull (see Figure 3.40), or a sufficiently large position displacement over the last few state updates.
For a time step Δt the process function becomes

$$\mathbf{x}_k^- = \begin{pmatrix} \mathbf{P}^W \\ \dot{\mathbf{P}}^W \end{pmatrix}_k^- = \begin{cases} \mathbf{x}_{k-1}^+ & \text{upon initialization, and} \\ \begin{pmatrix} 1 & \Delta t \\ 0 & 1 \end{pmatrix} \mathbf{x}_{k-1}^+ & \text{after apparent motion was observed.} \end{cases} \tag{3.43}$$

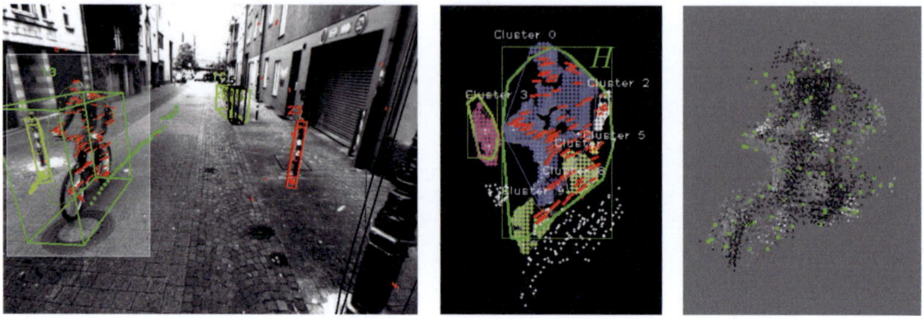

Figure 3.40: Left: Input image with final aligned 3D bounding box $\mathbf{B}$. **Middle:** Disparity clusters $\mathbb{S}$ generated by the cyclist, projected object hull H and feature flow (red lines). **Right:** Object shape $\mathbf{O}$ accumulated over 20 state updates, and downsampled shape information (green points) used to determine the hull H.

To incorporate the detections, the observation model needs to map the position $\mathbf{P}^W$ to image space, where segments are measured. This transformation is given as

$$\begin{pmatrix} u \\ v \\ \delta \end{pmatrix} = h\left(\mathbf{P}^W\right) = F^{-1}\left(T_W \mathbf{P}^W\right) = F^{-1}\left(\begin{bmatrix} x \\ y \\ z \end{bmatrix}\right) = \begin{pmatrix} \frac{xf}{z} + c_u \\ \frac{yf}{z} + c_v \\ \frac{Bf}{z} \end{pmatrix} \tag{3.44}$$

The Kalman update is carried out with μ_G and its covariance Σ_G to find the new state estimate $\mathbf{x}_k^+$.
The object shape $\mathbf{O}$ consists of the 3D points of the past detections and is transformed along with each position update. In order to further compress the environmental information, the object shape is abstracted to a minimal three-dimensional aligned bounding box model $\mathbf{B} = (w, h, l, \alpha)$. The box is centred on the state position $\mathbf{P}^W$ and defined by three dimension parameters and one alignment angle. The parameters are measured from the shape $\mathbf{O}$. A principal component analysis is applied to find the current alignment and estimate the box dimensions along the principal components.

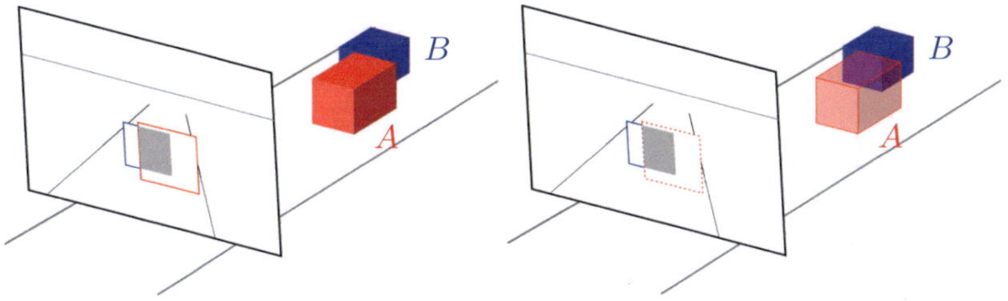

Figure 3.41: **(a)** Object A occludes Object B. The criterion is the overlap of the projected hulls. **(b)** The missed detection of object A allows the conclusion that track A does not exist anymore, since it would prevent the detection of B.

3.2.5 Object Track Management

A robust multi-object tracking system requires a logic that maintains the object tracks. It is responsible for track initialization and track deletion. Detections which cannot be explained by existing tracks should lead to the initialization of new tracks. Existing tracks which cannot be confirmed with current measurements should be deleted.

New tracks are initialized with segments which could not be assigned to existing tracks. More considerations are required to terminate existing tracks. Intuitively, a track should be deleted after the object could not be detected for a certain time. Often however, it can be explained why an object was currently not detected. The object can have left the sensor range, or it can be occluded by another object. In both cases even a series of missing detections is not a criterion to terminate a track. In other words, an object can only be deleted if it is currently measurable, i.e. within the detection range of the sensor and also visible.

Reasoning about the measurability means reasoning about the occlusion state of an object. The criterion used here is based on the projected object hulls H in image space. For object A to occlude another object B, the conditions are: 1. A is located closer to the camera than B. 2. Object hull B is covered by A to a certain amount $\frac{H_A \cap H_B}{H_B} > \epsilon$ (Figure 3.41a). In a similar way overlap with the visible camera range can be determined.

Object tracks are terminated and deleted from the model based on two possible conditions: 1. an object is measurable but could not be detected for a certain time. 2. an object is measurable but was not detected, and it occludes another detected object. See Figure 3.41b for an example. Object A occludes object B in this scenario. One can expect to detect object A, while object B is not measurable. If in this situation object B is detected, and object A remains undetected, the conclusion can be drawn that object A does not exist anymore – it would prevent the detection of B. The track of object A is consequently terminated.

3.2.6 Evaluation

The introduced method recognizes the presence of objects and approximates their dimensions, and their movement. This information is immediately useful for targeted close range navigation with avoidance of static obstacles and collision with moving objects. Beyond detection, these tasks require stable tracking of the surrounding objects. Strong noise in the underlying depth estimation and frequent inter-object occlusions give rise to a large set of possible failure sources that can manifest in state estimation errors, undetected objects, or tracking loss.

Clearly, the overall accuracy and precision of such system is a function of the sensor to object distance. The tracking scheme includes a loop, in which the current detections are partly influenced by the existing tracks. On one hand this facilitates a low-level sensor data segmentation without prior knowledge about the objects, on the other hand it poses the risk to amplify early made estimation errors over long periods of time. Non-occluded objects enter the scene at the limit of sensor perception range. During forward motion with small turn rates, static objects usually emerge at the sensor distance limit rather than at the lateral sensing limit. Thus, new objects are mostly initialized at the point of greatest measurement uncertainty. Decisions made at this point are carried on until the object track is deleted from the model. A typical such decision can be a merge of two pedestrians into one object.

The maximal range of reliable perception depends on the sensed image resolution. The sensed resolution contributes linearly to the algorithms runtime (low level segmentation) and limits the possible measurement frequency.

Reliable data association of moving objects across sensor readings requires a measurement frequency as high as possible. Thus, the system design needs to balance perceived distance against the ability to handle objects with increased relative velocities.
To ensure a measurement frequency of at least 10 Hz, the input disparity data is downsampled by factor four in our setting, resulting in an effective resolution of 160×120 image points. For reliable object initialization a minimal segment size of 5 image points is requested which allows to detect small poles in around 15 m distance. Assuming a moderate ego-velocity of $5\,\mathrm{km\,h^{-1}}$ the time to collision with such obstacles amounts to 10 s and allows for a targeted circumnavigation.
To evaluate the performance of the algorithm quantitatively, we investigate the number of wrong object detections (false positives) and missed object detections (false negatives). False positives occur mainly as consequence of loosing track of a moving object. The lost track is kept in the model until it becomes deleted after a number of missing detections. Other reasons for false positives are single objects that are modelled with multiple, partly overlapping tracks. False negatives result mainly from inaccurate scene background estimation (Section 3.1). The data that is fed into the algorithm is assumed to belong to scene foreground solely. Consequently, an erroneous scene background partitioning propagates as error into the object tracking. Scene parts that are concealed behind false positive (non-existing) vertical structures are disregarded and lead to false negative (missed) object detections. Undetected vertical structures on the other hand lead to false positive objects, for instance on building walls.
In Figures 3.42, 3.44, 3.45 and 3.46 we show results of the algorithm in different scenarios and camera setups. The first sequence is captured using our head-worn helmet setup (Section 4.1) in a typical urban street canyon scenario with many static scene elements (cars, signs, plants, close walls) and some moving objects like cars and cyclists. The second and third sequence originate from the series of work of Ess et al. [23] and are captured from a stroller in populated pedestrian zones with many moving persons but less static objects. Compared to our own data, the camera motion is substantially smaller and smoother in this setting. Finally in Figure 3.46 qualitative results

Figure 3.42: Results on the first sequence, captured with a head-worn camera in a typical urban street canyon scenario.

from an automotive setup are shown. To assess the performance, true positive, false positive and negative detections as well as identity (id-)switches are counted frame-wise in the visible perception range with a limit of 15 m for the first two sequences.

Figure 3.42 shows results of the first sequence. It consists of 2147 frames, which contain 6783 possible object detections. The second image (left to right, top to bottom) shows missed detections due to a false positive vertical surface that spans over foreground objects (third image), the fourth image

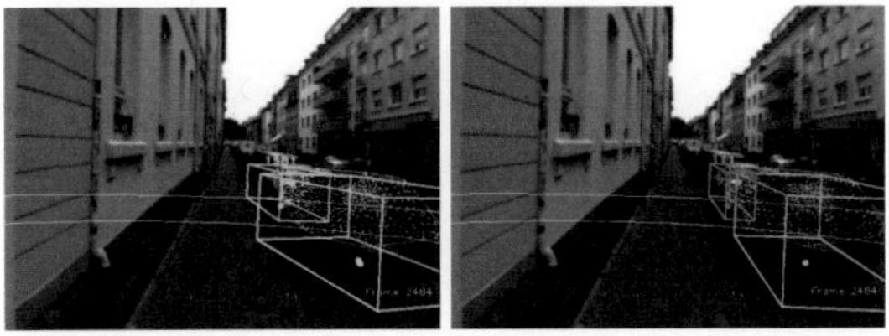

Figure 3.43: The alignment of obstacle bounding boxes can be inaccurate due to occluded parts of the objects (left). Taking scene context knowledge in form of building orientation into account, these errors can be mitigated. In the right image the frontal vanishing point is used for box alignment.

contains a false detection due to the missed background structure and a false object in the foreground (cyclist that was leaving the perception range). Since objects that have left the view are not removed from the model, the car (id 10) can be recognised in the fifth image after around 400 frames of non-visibility. In image 7 the long track of the cyclist becomes temporally merged with the signpost. The last 5 images show examples of different tracked moving objects with the ability to handle occlusion. Of all 6783 possible detections, 649 are missed (425 due to errors in background estimation). 1985 objects are falsely reported, 928 of these lie on building facades due to missed detections of background structure and are considered false positives. Overall an average recall of 90% at a precision of 76% is achieved. If background estimation errors are neglected these numbers raise to 96% recall at a precision of 86%. During the sequence 2 id-switches occur, where tracks switch onto another object.
In Figure 3.42 the current object shape models are contained (**O** projected as dots with bounding hull), in addition to the more abstract and compact aligned bounding box representation. The estimated box orientation often appears inaccurate for larger objects. Principally, only one side of the objects is visible. The measured "L"-shapes e.g. of cars lead to skewed principal components. Not only does it affect the box orientation, but it also leads to an erroneous dimension estimation. In structured urban environments one can exploit the typical alignment of objects with the surrounding geometric background scene structure, that is depicted accurately by the vanishing

Figure 3.44: Results on the second sequence *Jelmoli* [23] which is captured from a stroller.

directions. Replacing the principal component orientation estimation with the direction of the frontal vanishing point can mitigate this problem as shown in Figure 3.43.

From the second dataset shown in Figure 3.44 we quantitatively evaluate 500 frames in which 2243 possible detections occur. The upper 6 images show some typical tracks with merged objects. Since no concept knowledge is available, these tracks cannot be considered as erroneous detections (person with stroller, parents with child, two persons walking close to each other). The lower 6 images show some failure cases. A car partly enters the perception range due to camera rotation and initializes a false positive object. The child and the adult switch identity in the second image. Their tracks get lost upon leaving the perception range and cause false positive detections in the third image, besides two false negative detections (undetected poles on the right). The last two images show false positive detections in background clutter, a person in front tracked by two tracks (false positive), but also two long true positive tracks (purple) through full occlusion. Of the 2243 possible detections 205 are missed while 204 false detections are reported and 4 object tracks switch identity. It results in an averaged recall of 92% with a precision of 91%.

In Figure 3.45 results of the third sequence captured in a pedestrian zone are shown. Selected subsequences demonstrate the ability to create long consistent tracks also in dynamic environments with frequent full inter-object occlusion.

In Figure 3.46 a completely different camera setup was tested, as it is typically found in automotive platforms. Due to the higher image resolution and wider camera baseline the perception range could be increased to 30 m. The approach works well for small static obstacles and moderately moving traffic participants as cyclists. The high relative velocities of cars, coupled with the lower frame rate of 10 Hz in this dataset, require adjustments to the filter and the data association step. Particularly, the optical feature flow should be considered during the object prediction step. While it is unreliable for slowly moving objects, it should be a supportive clue here.

3.2.7 Discussion

The evaluation shows that the method is well able to detect generic objects with small amounts of missed detections and false alarms. Evaluated from the viewpoint of an object tracking system, a missed detection does not imply that the system is unaware of the object. It can simply be the result of delayed track initialization. Object handling, e.g. for motion planning, is a continuous problem in which the objects and their properties are the essence. Detecting the objects is only the intermediate step to learn about the objects, estimate their geometric appearance and their behaviour over time. Regarding the

Figure 3.45: Results on the third sequence *Bahnhof* [23].

object detection metrics, the tracking scheme that is applied here leads to an increase in recall compared to an independent detector, since missed detections can be bridged. On the other hand, it also leads to a decrease in precision, since objects that get lost during tracking are kept as false detection for many frames until they are eventually removed from the model.

As important as detecting objects is the ability to correctly estimate their motion. Only this allows to predict the surrounding situation, which is necessary to plan in dynamic environments. In this work, objects are described by their shape in form of an unorganized set of 3D points and a constant velocity assumption. This simple modelling provides the means to track objects over long times, including situations in which they temporarily leave the camera perception range, or in situations of full inter-object occlusion. The rough object shape estimation makes the state of occlusion observable. The estimated motion state allows to predict the object position over the period of missing measurements. This enables to proceed without explicit object detectors.

In the three typical scenarios there is no object within the 15 m range that remains completely unseen by the system over the whole sequence. This would be a requirement when used for collision avoidance with the static environment. The low number of id switches indicates accurate tracking also for moving objects. As argued earlier, a perception range of 15 m allows a collision time of 10 s for static obstacles at a moderate moving speed of 5 km h^{-1}. It can easily be assessed, however, that planning to cross a road in highly dynamic situations with cars – even when assuming moderate speeds of 30 km h^{-1} – requires a perception range of at least twice the current length. The perception limit of the system is mainly due to the strongly downsampled input data (disparity map by factor 2, foreground disparity points by factor 4). This hinders the reliable generation of object hypotheses in a distance greater than 20 m. Due to the small baselines, disparity gradients become very small and lead to more frequent undersegmentation in the early low-level stages of the algorithm. These propagate as merged objects over time and are currently not resolved.

The approach is not limited to our specific setup. We demonstrate this by evaluating datasets from different platforms with different viewpoints, different cameras, different baselines and different types of ego motion. Even the deployment in automotive settings is feasible as shown in our experiments in Figure 3.46, though few chosen parameters need further adjustment to the increased sensor resolution and viewing range.

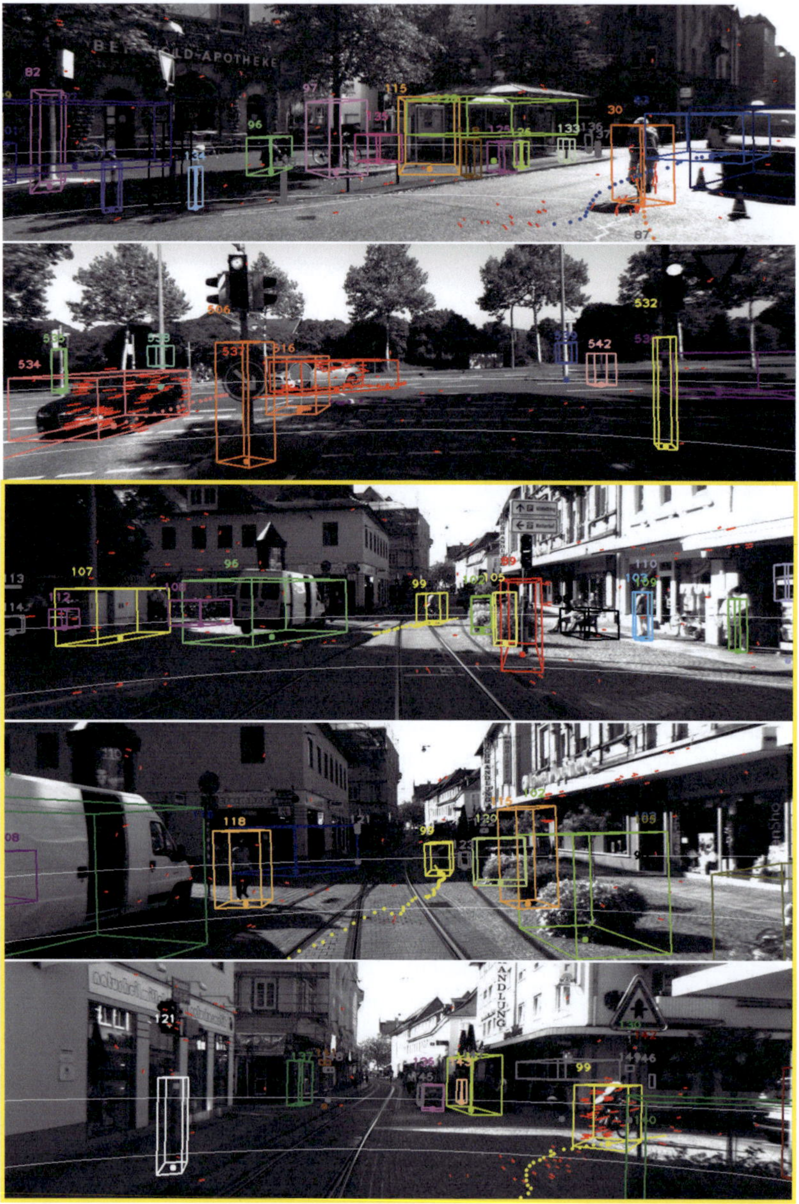

Figure 3.46: Some example scenes from a binocular camera in an automotive setup (part of the Kitti benchmark suite [27]). The perception range was increased to 30 m here. Besides traffic participants like cars, cyclists and pedestrians, also small obstacles like poles, bushes and signposts are recognized. The last two pictures show a comparison to related work of [68], in which semantic labels are used to filter objects from background.

3.2.8 Related and Future Work

The proposed approach resides among few other works of generic object tracking, that do not exploit concept knowledge about e.g. shape or appearance. Though binocular depth data is found in many applications for e.g. pedestrian detection, it is usually accompanied by object specific detectors. The depth data is applied to generate regions of interest to improve and disambiguate the detections. The sequences 2 and 3 were used in related work of Ess et al., e.g. [23], who use an appearance based detector to generate initial pedestrian hypotheses. Their reported performance shows especially a lower recall, that is likely explained by the fact that it explicitly relates to pedestrian detection, where groups of pedestrians that are merged into a single track count as false positive detections. Other approaches that build upon binocular depth data segmentation to generate detections include [3],[63] and [68]. All use grid maps in Euclidean space and segment by mode seeking, instead of directly segmenting the disparity map. The segmentation is followed by different strategies to discard background clutter. [3] and [63] eventually aim at pedestrian detection and verify segments with geometric properties or knowledge about appearance. [68] uses a pixel-wise semantic labeling that was trained to separate objects of arbitrary kind from background structure. The idea follows the recent trend of using object proposal methods [41] (which classify image regions based on a generic *objectiveness* criterion) to guide and speed up sliding window detectors. This allows to reduce the depth data to relevant object structure.

The most evident inaccuracies of the proposed approach are objects that sometimes seem too greedily merged, as for instance two persons walking next too each other. These cases almost always originate from object initialization in distance, where the low data resolution does not yet provide evidence to start multiple separate tracks. An object model is build that consists of two persons, which is tracked as such, as long the persons' paths do not diverge. It seems questionable whether two pedestrians walking close together constitute a valid object. There is, however, no way to tackle such problem without incorporating concept specific a-priori knowledge whatsoever, be it appearance, size, or shape to name a few. There is an obvious limit of what can be reached without knowing about the meaning of objects. A logical

extension of this work should hence make use of top-down reasoning, for instance by incorporating the promising upcoming results of appearance based semantic scene labeling, as was to some extent demonstrated by [68]. Currently however, the hardware required for online application of these algorithms exceeds our targeted platforms.
Another worthwhile extension might target the crucial and important measurement association step. In this work immediate, hard decisions about assigning measurements to object tracks are taken. A more robust strategy could delay the assignment and accumulate more evidence before decisions are made. Respectively, decide not only based on the current measurements but jointly decide for all measurements in a past timeframe. These kind of multiple-hypotheses tracking (MHT) algorithms surely have potential to dissolve a few mistaken assignments that lead to id switches or mistakenly instantiated object tracks. The chances come at the cost of a massive increase in computational complexity. Carefully chosen heuristic strategies to prune hypotheses are required in order to counteract the combinatorial explosion of assignments. In order to avoid hard assignment decisions, joint probabilistic data association (JPDA) filters are often applied. For each combination of measurement and object track the probabilities are estimated that the measurement originates from the object. During state estimation, each object track becomes updated with all measurements, weighted according to these probabilities. The standard JPDA framework assumes knowledge about the number of tracked objects, and further that each object generates at most one measurement per timestep – assumptions that both need to be overcome in our setting. Increasing interest growth around methods based on finite set statistics [33]. The mathematical framework provides statistical means to cast multi-object, possibly multi-sensor tracking into a single-sensor single-object Bayesian filter problem. Thereby, unifying the estimation of object number and their states in a single procedure where object appearance and disappearance become modelled within the state time transition. Sensors, objects, and the measurements they generate are modelled as random finite sets (RFS), sets of random variables with random cardinalities. This inherently allows for multiple measurements per object without the requirement of an explicit data association, which makes it interesting for extended object tracking

problems (e.g. [7]). However, as pointed out earlier, the amount of possible measurement sets quickly grows to an intractable number and necessitates suitable approximations of probable sets. In this work the partitioning of sets was approximated as a clustering problem, using the current object positions and their uncertainties as a clue. As such it could provide a starting point to embed the problem into the computational feasible approximations of RFS tracking algorithms, including the probability hypothesis density (PHD) filter, the Cardinalized PHD (CPHD) filter, or multi-Bernoulli (MB) filters. Eventually though, the performance of any such filter will be limited by the correctness of required assumptions regarding the environmental and sensor statistics. Not only are figures like object birth or false alarm rate hard to quantify, in mobile systems they are also highly dependent on the ever changing surrounding situation. This renders tracking of multiple generic extended objects a challenging problem that involves way more than following a signal through clutter.

4 Experimental Platform

4.1 Experimental Setup

Two wearable setups were build to test the system experimentally. Their basis is a binocular setup consisting of two PointGrey FL2-14S3M Firewire cameras, and Toshiba BU238 USB-3.0 cameras respectively. Kowa lenses with a focal length of 3.5 mm result in an opening angle of approximately 65° horizontally and 45° vertically. Images are captured monochrome with 640x480 pixel at 30 frames per second, synchronized via the Firewire/USB-Bus. Both setups are further equipped with an inertial measurement unit (IMU), which provides a ground truth of the camera orientation. The computational platform is a notebook that can be carried in a backpack. For disparity estimation we apply the semi-global matching [39] implementation of OpenCV[1] at half-resolution (320x240 pixel). Visual odometry is calculated using the libViso2[2] implementation [30].

The first setup, that is shown in Figure 4.1, was used to record the datasets used throughout this work. It uses a bicycle helmet as platform, in which all sensors are flush-mounted. The camera baseline amounts to 18 cm. It allows to record data from a natural viewpoint without interfering normal movement. The captured data is representative for first-person (egocentric) video understanding applications. The embedded IMU is an Xsens MTi-300 which is drift compensated using gravity and the earth's magnetic field. In scenarios of short translational accelerations the orientation estimate of this MEMS sensor based unit is a suitable reference.

[1] http://opencv.org

[2] http://www.cvlibs.net/software/libviso/

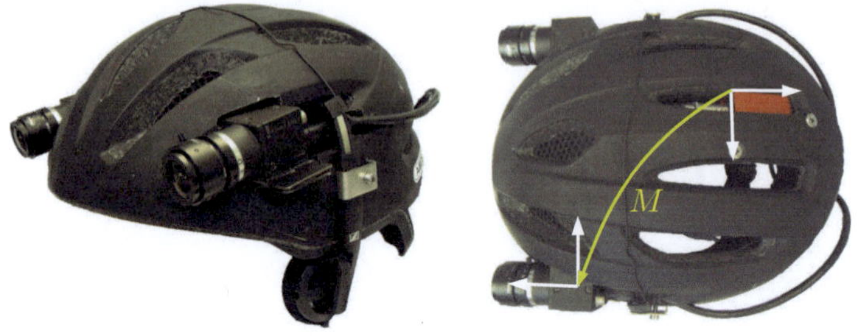

Figure 4.1: Experimental setup used to record data from an egocentric viewpoint. It is equipped with cameras, an IMU (orange) and attached headphones.

The second setup in Figure 4.3 is a demonstrator for virtual reality applications. The cameras are mounted sideways onto an Oculus Rift stereoscopic head-mounted display (HMD), such that they point into the current viewing direction of the user. The HMD contains an embedded IMU, which is used by virtual reality applications to determine the head orientation.

Sensor Calibration The cameras are calibrated intrinsically using a pinhole model as well as extrinsically with respect to each other with the one shot solution of [28], which uses multiple chessboard patterns as calibration targets. The orientation between camera rig and the IMU is given by the rigid transformation $M \in SO(3)$. Since the IMU is solely used as orientation reference, the translational component can be neglected here without any loss. After gathering corresponding delta orientations of IMU ΔR_{imu_i} and camera (visual odometry) ΔR_{cam_i} we find M by minimizing the Frobenius norm

$$\min_{M(\theta,\varphi,\rho)} \sum_i ||M \; \Delta R_{imu_i} M^{\mathrm{T}} - \Delta R_{cam_i}||_F. \tag{4.1}$$

The delta orientations need to cover all three degrees of freedom, which can easily be accomplished with the small sized setup.
Assuming error-free inertial measurements, an estimate of the visual odometry errors can be made. Comparing frame to frame delta poses a standard deviation of $\sigma_R \approx \; 0.2°$ around all three axis can be observed, at small biases in the range of $10^{-3}°$. Since the IMU does not provide translational

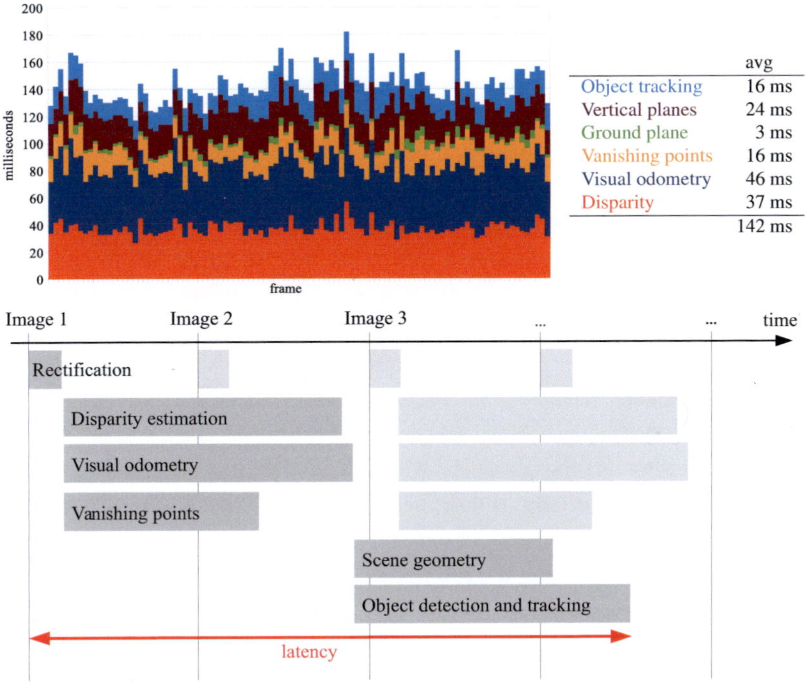

Figure 4.2: Top: Individual processing times of the main tasks over 100 frames on a dual core 2.4 GHz CPU. Disparity estimation and visual odometry account for more than half of the processing time. **Bottom:** Parallel processing and pipelining of modules leads to a schedule with smaller latency and higher throughput.

information, we cannot use it to estimate the visual odometry translational errors. Instead, we assess this error by comparing delta poses of the position referenced setup used in the KITTI benchmark [27] ground truth to find an estimate of $\sigma_{\mathrm{t}} \approx 0.05$ m.

4.1.1 Software Framework

The overall perception process can be broken down into a small number of different tasks, some of which are independent of each other, some of which are not. These main tasks are disparity estimation, feature matching and motion estimation (visual odometry), line extraction and vanishing point estimation, geometry estimation, and object detection and tracking. These tasks need to be embedded into a software architecture that minimizes the overall latency between a new input image and the high-level output of foreground objects and scene geometry, while maximizing the throughput of individual tasks in terms of processed images per time. This is crucial particularly in case of the egomotion estimation task, which depends on a high throughput to ensure camera pose tracking also under fast motion.

To fulfil these requirements, the framework is implemented as a blackboard architecture. Each task is processed by an individual module. All modules communicate with each other via a common storage (blackboard). A module starts processing, whenever all required information are available in the storage and stores its processed and enriched information back to the storage. This in turn triggers other modules, which depend on this information, to start their task. The scheduling of tasks arises implicitly by defining the module dependencies. Modules work in parallel whenever possible. This ensures a small latency by exploiting the full potential of current multi-core processing units. Modules process data preemptive, which leads to a pipelining effect that ensures a high overall throughput. Individual module times are shown in Figure 4.2 (top) over 100 input frames. Sequentially processing all modules would cause an average latency of 142 ms, or a framerate of 8 Hz. In practice, disparity estimation, visual odometry and vanishing points do not depend on each other and can be processed in parallel. Obstacle detection requires the background scene geometry, thus both should be run sequentially. To parallel these tasks nonetheless, the obstacle detector uses a prediction of the background geometry with the current egopose. A schematic of the resulting module schedule is visualized in Figure 4.2 (bottom). The effective latency for the obstacle output is reduced to 62 ms, which corresponds to a framerate of 15 Hz.

The time-wise last module in the process chain is an information sink that provides the scene geometry, obstacles, and the egopose over a network connection. Arbitrary client applications can connect to this server to receive the modelled environment information. Experiments with two such applications are described in the following.

Figure 4.3: Hardware setup and substitution of visual scene perception as used in this virtual reality experiment.

4.2 Virtual Reality Experiment

If solely the abstract environment model was available to a human operator, could he or she still successfully plan a path, negotiate obstacles, or navigate towards objects? This question inspired a virtual reality experiment, in which the person is provided with a three-dimensional visual rendering of the modelled data in the head-mounted stereoscopic display. The side mounted cameras point into the viewing direction such that they perceive information similar to the human visual sense. The user's visual perception is substituted by algorithms that detect objects and model scene geometry. Figure 4.3 shows an impression of the virtual view.

By coupling the system's feedback with a human operator a special loop is created, in which the system's output influences the immediate behaviour of the user, which in turn affects the future input of the system. The influence of such loop can not be considered in offline experiments with recorded data. A critical issue of this coupling is the problem of latency. Human perception immediately adapts to changes in the visual field caused by the own action, e.g. a rotation of the head. Delaying or suppressing the visual impressions causes a conflict between action and perception that leads to the common problem of motion sickness. To avoid motion sickness, latencies of a few milliseconds must not be exceeded, which can hardly be achieved by a sequential computer vision process. Even though we have been able to reduce the latency to around 60 ms by exploiting parallel computing, it remains a magnitude to high. A solution is to predict the perception based on the environment model. The viewing direction is available with smaller latency from visual odometry, or almost free of latency when using the IMU. The user is constantly provided with (maximally 60 ms) aged data, which is, however, perceived correctly with negligible delay. Furthermore, it allows to increase the refresh rate of given feedback independently of the processing time of all underlying perception algorithms.

Figure 4.4: Results of the virtual reality experiment. The colored overlay visualizes the path which the participants took while navigating from tree to tree back to the starting point.

Figure 4.4 shows the setting of the conducted experiment. Participants started on the right hand side into a given initial starting direction. Their task was to navigate from tree to tree in a circle until they reached their rough initial starting position, which was marked within the visualization. The scene was filmed from the viewpoint in Figure 4.4 in order to record and overlay the paths walked by the participants.
Some conclusions can be drawn from these experiments. The walked paths show that the modelled object information principally suffices to navigate towards and around objects in the range of 15 m. Compared to the visual impression that the human perception provides, the artificial visualization is very sparse. This concerns especially the medium distant scene background, which usually provides important orientation clues, but is largely absent here. This leads to a quickly settling feeling of disorientation, even though the representation itself is consistent. The final offset between starting and finishing position gives an idea about the accumulated position drift during the time of walking. To mitigate the orientation drift, the IMU was used to correct the visual odometry estimate in this experiment.

4.3 Application to Assist Visually Impaired Persons

The virtual reality experiment indicates the usefulness of the abstract information for a human user during medium range navigation and obstacle avoidance tasks. The human visual perception, that particularly includes the interpretation and understanding of the vast amount of sensed visual information, was almost entirely replaced by the perception framework. The visual sense was merely used to transmit the abstracted information.
If one succeeds in substituting this visual feedback with another sensory modality, the perception framework can serve as a powerful technical assistance aid for visually impaired persons. The difficulty in substituting the visual sensory channel is its immense capacity. The reasonable alternatives are the tactile and acoustic modalities. However, these are trained to sense information complementary to the visual world. Neuroscience research has shown the ability of the brain to perform a substitution of sensory modalities through other signals. A well known example is *The vOICe* system [61],

which aims to substitute the visual with the auditory sense. Camera images are directly mapped into sounds by a left-to-right scan through each single video frame, which the user perceives with a headphone. Similar studies have been conducted using tactile feedback [10]. After long periods of training the brain can adapt to these signals and transform them into visual perceptions. Many different technical prototypes have been developed to assist visually impaired persons in a more easily accessible way [11, 19, 112]. Such systems usually aim to inform the user about non-traversable directions in the scene (e.g. [74, 24]). Depending on the sensor this can require the user to actively scan the environment similar to a torch [8, 42]. In these systems the feedback can be directly coupled with the sensed information. In cases where a wide sensor range permits passive sensing, the design of intuitive feedback is hard, because the amount of information remains very high. The interpretation of the complex feedback signals is left to the user [32, 60, 10], and can cause substantial cognitive load. Another strategy is to lead the user into walk-able free space by solving a path planning problem [83, 71]. Intuitive tactile feedback can be given for instance with vibrating elements around a belt. However, the system takes high-level decisions for the user, which can constitute a serious barrier for usage.

The perception framework presented in this work can enable a solution in between these approaches. A system, that passively senses the environment, informs the user with intuitive feedback and does not take decisions but leaves the user in full control. The challenge consists in first, filtering the relevant information from the large amount of sensed data and second, providing it in a form that is intuitively understandable. The abstract representation of the framework provides the necessary basis. Rather than informing about the free space, the user can be specifically informed about potential obstacles, their kind and their behaviour.

In context of the OIWOB project [65], this was realized by augmenting the acoustic reality of visually impaired users with spatial sounds that are perceived through an open headphone. Each object emits a sound that the user perceives spatially as if a speaker was attached to it. This is achieved by binaural rendering, which creates sounds that can be correctly localized in direction and distance using a headphone [37]. It requires the (individual) head-related transfer function (HRTF) of both ears to render a sound depending on the direction and distance as if it was naturally distorted through the outer ear and delayed according to the ear distance [9]. The choice of sounds was made based on a number of simulator studies with visually impaired users. Besides informing about the location of objects, different sounds can encode semantic information, e.g. the object type. Sounds were chosen to be clearly distinguishable from natural sounds, while still inducing an association with common semantic concepts. Further design choices had to regard the localizability of chosen sounds and their harmony with respect to each other. A summary of these considerations and the conducted studies can be found in [106].

The system was prototypically realized using the helmet setup in Figure 4.1, and a notebook carried in a backpack. All tracked objects were additionally geometrically classified into the classes flat, pole-like, overhanging and dynamic, and a characteristic sound was assigned to each class. A field test with 8 visually impaired participants was conducted to confirm the principal function of such system.

Fieldtest The field test consisted of an artificial obstacle course (Figure 4.5, top row). The participants' task was to orientate themselves along the edge between lawn and pavement with the white cane, while avoiding different obstacles that were placed along the path, some directly on the edge, some to the left or right.

A familiarization with the virtual spatial sound sources could be observed for most participants within minutes of practice. A single obstacle is sufficient in this phase to experience the principal of scene fixed sounds that adjust according to the movement and head rotations. To keep the cognitive load of untrained users within a reasonable limit, the three most relevant objects in terms of distance and viewing direction were selected and converted into sounds. This can include objects which are located lateral, outside the visible camera range. Therefore, stable object tracking also under temporary invisibility is a prerequisite. In the beginning of the parcours, the users tended to stop walking whenever a new obstacle was sonified and turn their head in order to confirm the sound direction. Later they slowed down their walking speed until the obstacle was in reach of the white cane. The obstacle distance was perceived by sound volume, which turned out to be very difficult to judge. A longer training phase is required to learn the relation between sound volume and distance. Other options are conceivable, for example an intermittent sound presentation that encodes distance by the frequency of sound pulses. Those participants, who are independent and mobile on their everyday routes, could achieve an obstacle avoidance behaviour in a second or third walk through the parcours. Two participants who felt instantly very comfortable with the system were taken on a free walk on the sidewalk of a nearby road. They communicated their impressions verbally. Besides avoiding collisions, they were also able to follow the sound that a guiding person virtually emitted. Remaining difficulties were caused by the unknown object extensions. Though it is modelled by the perception system, the size of objects is currently not reflected in the feedback. It remains an open question how the size of objects sounds, mainly because it is a feature out of the usual scope of hearing.

Figure 4.5: Impressions of the conducted field tests with visually impaired people.

Many extensions to this prototype are imaginable. The framework's environment model contains additional information that could be helpful for a visually impaired user, and also the potential of acoustic feedback is by far not exhausted. In real environments, sound reflections on surfaces give impressions of space, and Doppler effects indicate the motion of objects. Furthermore, the feedback could include less concrete information such as the direction of vanishing points, which might be a valuable orientation clue. The versatile camera-based setup allows to extend and refine the environment model to particular demands of visually impaired people. Detecting crosswalks, bus stops, signs or recognizing text are just a few examples of expressed wishes.

5 Conclusion

In this work a framework was presented that enables a mobile system to perceive its environment with a binocular camera. In order to operate in unknown environments these systems need to understand their surrounding in a spatial as well as temporal manner. To facilitate common, yet unspecific tasks like targeted navigation, interaction or obstacle avoidance, methods were presented that abstract the raw camera data into a semantically enriched meta representation. It models the environment on a semantic level of object instances. This forms a basis for higher level reasoning tasks and leads to a very compact environment representation that is well suited for the constrained resources of a mobile platform.
These opportunities are accompanied by various technical challenges. Cognitive object detection and recognition processes are to a great part driven by top down reasoning from known concepts to sensory features. Such knowledge is not available unless it is explicitly modelled or learned. Thus, it needs to be substituted by other clues. The technical possibilities of mobile systems are limited, not only in terms of computational power but also in terms of sensor coverage and accuracy. Perception algorithms have to cope with low resolution sensor data, have to be robust to erroneous measurements and have to be efficient enough to allow for real-time operation.
Different algorithms were introduced that, taken in combination, solve this problem. We first argued that typical environments of mobile systems can be decomposed into a static scene background structure which is filled by independent, possibly moving objects in the scene foreground. The scene background structure was modelled with geometric planes which express the ground around the system, and typical surfaces like building facades, fences or bushes. Estimating and tracking these surfaces is difficult because of their large variability, their large distances and large depth uncertainties in the binocular reconstruction. As a complementary mid-level feature the

concept of vanishing points was introduced. An algorithm was developed to estimate and track an accurate model of scene directions using a sequence of past images. The estimated directions were exploited in different ways. The direction of vanishing points enabled us to keep track of the ground surface also during major occlusions and temporal invisibilities due to the surface being out of view. Other than in most related systems, permanent ground visibility cannot be assumed due to the unconstrained camera setup. Based on the ground orientation we developed a method to estimate the background structure that delimits the scene. It is represented by geometric planes which are vertically aligned with the floor. Afterwards, a geometric representation was found for staircases. Image edge observations and vanishing directions were used to develop an efficient algorithm that models stairs with a minimal set of geometric parameters. Stair steps are tracked over time in order to enable the traversal of the staircase.
Pose estimation and scene modelling were treated independently in this work. The environment model is locally consistent but is globally afflicted with the drift of the visual odometry. In order to mitigate this drift, we showed how scene vanishing directions can be fed back into the pose estimation. This leads to an environment model that is consistent with the surrounding scene. The scene background model finally enabled to separate foreground objects from the scene. A method was proposed to track multiple unknown objects in low resolution disparity measurements. For each object we estimated the direction of movement, the velocity, and a shape model, which enabled to track objects through full occlusions. Experiments on challenging inner urban datasets showed that the methods are able to handle cluttered scenes with many independently moving objects and major occlusions. The perception range of 15 meters allows for targeted path planning under moderate ego-velocities. It could easily be extended on less limiting platforms, where wider binocular baselines are possible, or image resolution can be increased. This was demonstrated on data from an automotive platform. The object tracking itself produces a very small object miss-rate. Most errors occur due to inaccuracies of the background model in the limit of the perception range and are uncritical in terms of collision avoidance.

Foreground objects, scene background model and the relative camera pose are the output of the framework. Arbitrary applications can build upon this meta representation of the environment. We demonstrated two applications related to wearable systems, in which the cameras are worn on a persons' head. The human perception was largely replaced by a visual or acoustical rendering of the modelled information. These applications are particularly challenging because of their highly dynamic camera motion. The system output influences the behaviour of the user, and thereby the future input of the system. The visual feedback resembled the natural three-dimensional visual impression that a sighted person has, albeit reduced to a minimum. In the acoustical feedback, object locations were indicated by spatial, binaural sounds. This form of feedback has great potentials in the assistance of visually impaired people. Both experiments could show that the modelled environment information is sufficient to move towards and around obstacles. A targeted navigation becomes possible in the current perception range of 15 meters. This was widely independent of the feedback modality. For sighted persons the visual feedback was very clear and obvious. Visually impaired users were able to make use of the acoustic environment augmentation after few minutes of practice. This intuitive kind of feedback was made possible by the high degree of sensor data abstraction. The semantic level of information used here is currently unique in the field of technical aids for visually impaired people. The increased perception range allows to sense potential dangers earlier and avoid them in a targeted way. Moreover, overhanging objects are detected which pose a particular danger since they cannot be sensed with the white cane. As such, a system like this could contribute to the safety and independent mobility of visually impaired people. The versatile camera setup allows for many potential functional extensions, some of which were discussed in Section 4.3.

To make a system like this suitable for daily use, few user-oriented adaptations are conceivable. A small sized binocular camera system embedded into a light pair of glasses is feasible. A practical issue is the calibration of the camera setup. A onetime calibration requires a torsionally rigid frame, which contradicts the wish for unobtrusiveness. Instead, an online calibration method should be embedded to continuously accommodate at least the extrinsic camera alignment. None of our proposed algorithms relies on highly parallelized graphical processing units. With careful adjustments the computational power of a state-of-the-art smartphone could already be sufficient as computational platform. In a recent project, Google Tango [34] demonstrates spatial sensing for indoor scenarios on hand-held devices, which includes localization and depth sensing. These are essential parts of the proposed framework and account for more than half of the workload (Figure 4.2). An efficient implementation of these modules for the architecture of mobile devices would be a requirement but appears in very close reach.

Some future directions and possible algorithmic extensions were discussed in Sections 3.1.3.4 and 3.2.7. Other extensions of this work could aim at special application scenarios. The proposed environment representation is well suited for close range navigation tasks. Objects can be used as navigation targets, and background geometry gives valuable clues about the general scene alignment. Scenes contain more features which could be helpful here. Consider for instance guidelines like curbstones, the route of a pedestrian path through a garden, or crosswalks in traffic scenarios. The floor condition (e.g. lawn, flowerbed, water) can be a valuable feature to judge about traversable and impassable areas. A scene background model for an autonomous vehicle should be extended by a model of the road and its lanes. All these extensions require to provide more specific domain knowledge than was done in this work. Our only assumption was a flat world, that is delimited by vertical surfaces and populated with movable boxes.

Bibliography

[1] C. Akinlar and C. Topal, "Edlines: A real-time line segment detector with a false detection control," *Pattern Recognition Letters*, vol. 32, no. 13, pp. 1633–1642, 2011.

[2] M. Antone and S. Teller, "Automatic recovery of relative camera rotations for urban scenes," in *IEEE Conference on Computer Vision and Pattern Recognition (CVPR)*, 2000, pp. 282–289.

[3] M. Bajracharya, B. Moghaddam, A. Howard, S. Brennan, and L. H. Matthies, "A fast stereo-based system for detecting and tracking pedestrians from a moving vehicle," *International Journal of Robotics Research*, vol. 28, no. 11-12, pp. 1466–1485, 2009.

[4] O. Barinova, V. Lempitsky, E. Tretiak, and P. Kohli, "Geometric image parsing in man-made environments," in *European Conference on Computer Vision (ECCV)*, 2010, pp. 57–70.

[5] S. T. Barnard, "Interpreting perspective images," *Artificial Intelligence*, vol. 21, no. 4, pp. 435–462, 1983.

[6] J. C. Bazin, Y. Seo, C. Demonceaux, P. Vasseur, K. Ikeuchi, I. Kweon, and M. Pollefeys, "Globally optimal line clustering and vanishing point estimation in manhattan world," in *IEEE Conference on Computer Vision and Pattern Recognition (CVPR)*, 2012, pp. 638–645.

[7] M. Beard, S. Reuter, K. Granström, B. T. Vo, B. N. Vo, and A. Scheel, "A generalised labelled multi-bernoulli filter for extended multi-target tracking," in *IEEE International Conference on Information Fusion (FUSION)*, 2015, pp. 991–998.

[8] J. M. Benjamin, "The new c-5 laser cane for the blind," in *Carnahan Conference on Electronic Prosthetics*, 1973, pp. 77–82.

[9] J. Blauert, *Spatial hearing : the psychophysics of human sound localization.* MIT Press, 1997.

[10] N. Bourbakis, "Sensing surrounding 3-d space for navigation of the blind," *Engineering in Medicine and Biology Magazine, IEEE*, vol. 27, no. 1, pp. 49–55, 2008.

[11] J. A. Brabyn, "New developments in mobility and orientation aids for the blind," *IEEE Transactions on Biomedical Engineering*, vol. BME-29, no. 4, pp. 285–289, 1982.

[12] F. Camposeco and M. Pollefeys, "Using vanishing points to improve visual-inertial odometry," in *IEEE International Conference on Robotics and Automation (ICRA)*, 2015, pp. 5219–5225.

[13] V. Chari and C. Jawahar, "Multiple plane tracking using unscented kalman filter," in *IEEE/RSJ International Conference on Intelligent Robots and Systems (IROS)*, 2010, pp. 2914–2919.

[14] J. Civera, D. Gálvez-López, L. Riazuelo, J. D. Tardós, and J. M. M. Montiel, "Towards semantic slam using a monocular camera," in *IEEE/RSJ International Conference on Intelligent Robots and Systems (IROS)*, 2011, pp. 1277–1284.

[15] N. Cornelis, B. Leibe, K. Cornelis, and L. V. Gool, "3d urban scene modeling integrating recognition and reconstruction." *International Journal of Computer Vision*, vol. 78, no. 2-3, pp. 121–141, 2008.

[16] J. Corso, D. Burschka, and G. Hager, "Direct plane tracking in stereo images for mobile navigation," in *IEEE International Conference on Robotics and Automation (ICRA)*, 2003, pp. 875–880.

[17] J. Coughlan and A. L. Yuille, "Manhattan world: compass direction from a single image by bayesian inference," in *IEEE International Conference on Computer Vision (ICCV)*, 1999, pp. 941–947.

[18] I. J. Cox, "A review of statistical data association techniques for motion correspondence," *International Journal of Computer Vision*, vol. 10, no. 1, pp. 53–66, 1993.

[19] D. Dakopoulos and N. Bourbakis, “Wearable obstacle avoidance electronic travel aids for blind: A survey,” *IEEE Transactions on Systems, Man, and Cybernetics*, vol. 40, no. 1, pp. 25–35, 2010.

[20] J. Delmerico, D. Baran, P. David, J. Ryde, and J. Corso, “Ascending stairway modeling from dense depth imagery for traversability analysis,” in *IEEE International Conference on Robotics and Automation (ICRA)*, 2013, pp. 2283–2290.

[21] P. Denis, J. H. Elder, and F. J. Estrada, “Efficient edge-based methods for estimating manhattan frames in urban imagery,” in *European Conference on Computer Vision (ECCV)*, 2008, pp. 197–210.

[22] W. Elloumi, S. Treuillet, and R. Leconge, “Tracking orthogonal vanishing points in video sequences for a reliable camera orientation in manhattan world,” in *International Congress on Image and Signal Processing (CISP)*, 2012, pp. 128–132.

[23] A. Ess, B. Leibe, K. Schindler, and L. Van Gool, “Moving obstacle detection in highly dynamic scenes,” in *IEEE International Conference on Robotics and Automation (ICRA)*, 2009, pp. 4451–4458.

[24] G. P. Fajarnes, L. Dunai, V. S. Praderas, and I. Dunai, “Casblip-a new cognitive object detection and orientation system for impaired people,” *trials*, vol. 1, no. 2, p. 3, 2010.

[25] K. Fukunaga and L. Hostetler, “The estimation of the gradient of a density function, with applications in pattern recognition,” *IEEE Transactions on Information Theory*, vol. 21, no. 1, pp. 32–40, 1975.

[26] A. Geiger, M. Lauer, and R. Urtasun, “A generative model for 3d urban scene understanding from movable platforms,” in *IEEE Conference on Computer Vision and Pattern Recognition (CVPR)*, 2011.

[27] A. Geiger, P. Lenz, and R. Urtasun, “Are we ready for autonomous driving? the kitti vision benchmark suite,” in *IEEE Conference on Computer Vision and Pattern Recognition (CVPR)*, 2012.

[28] A. Geiger, F. Moosmann, Ömer Car, and B. Schuster, "Automatic camera and range sensor calibration using a single shot," in *IEEE International Conference on Robotics and Automation (ICRA)*, 2012.

[29] A. Geiger, M. Roser, and R. Urtasun, "Efficient large-scale stereo matching," in *Asian Conference on Computer Vision (ACCV)*, 2010.

[30] A. Geiger, J. Ziegler, and C. Stiller, "Stereoscan: Dense 3d reconstruction in real-time," in *IEEE Intelligent Vehicle Symposium (IV)*, 2011.

[31] J. J. Gibson, *The perception of the visual world.* Houghton Mifflin, 1950.

[32] J. Gonzalez-Mora, A. Rodriguez-Hernandez, E. Burunat, F. Martin, and M. Castellano, "Seeing the world by hearing: Virtual acoustic space (vas) a new space perception system for blind people." in *Information and Communication Technologies (ICTTA)*, 2006, pp. 837–842.

[33] I. R. Goodman, R. P. Mahler, and H. T. Nguyen, *Mathematics of Data Fusion.* Norwell, MA, USA: Kluwer Academic Publishers, 1997.

[34] "Google tango," https://get.google.com/tango/, accessed February 2017.

[35] K. Granström and M. Baum, "Extended object tracking: Introduction, overview and applications," *CoRR*, vol. abs/1604.00970, 2016.

[36] A. Gupta, A. A. Efros, and M. Hebert, "Blocks world revisited: Image understanding using qualitative geometry and mechanics," in *European Conference on Computer Vision (ECCV)*, 2010.

[37] T. Hermann, A. Hunt, and J. G. Neuhoff, Eds., *The Sonification Handbook.* Logos Publishing House, 2011.

[38] D. Hernandez and K.-H. Jo, "Stairway segmentation using gabor filter and vanishing point," in *International Conference on Mechatronics and Automation (ICMA)*, 2011, pp. 1027–1032.

[39] H. Hirschmuller, “Stereo processing by semiglobal matching and mutual information,” *IEEE Transactions on Pattern Analysis and Machine Intelligence*, vol. 30, no. 2, pp. 328–341, 2008.

[40] D. Hoiem, A. A. Efros, and M. Hebert, “Recovering surface layout from an image,” *International Journal of Computer Vision*, vol. 75, pp. 151–172, 2007.

[41] J. Hosang, R. Benenson, P. Dollár, and B. Schiele, “What makes for effective detection proposals?” *IEEE Transactions on Pattern Analysis and Machine Intelligence*, 2015.

[42] A. Hub, J. Diepstraten, and T. Ertl, “Design and development of an indoor navigation and object identification system for the blind,” *SIGACCESS Access. Comput.*, no. 77-78, pp. 147–152, 2003.

[43] V. Huttunen and R. Piché, “A monocular camera gyroscope,” *Gyroscopy and Navigation*, vol. 3, no. 2, pp. 124–131, 2012.

[44] C. Kessler, C. Ascher, N. Frietsch, M. Weinmann, and G. Trommer, “Vision-based attitude estimation for indoor navigation using vanishing points and lines,” in *IEEE/ION Position Location and Navigation Symposium (PLANS)*, 2010, pp. 310–318.

[45] J. Kosecká and W. Zhang, “Video compass,” in *European Conference on Computer Vision (ECCV)*, 2002, pp. 476–490.

[46] T. Kroeger, D. Dai, and L. V. Gool, “Joint vanishing point extraction and tracking,” in *IEEE Conference on Computer Vision and Pattern Recognition (CVPR)*, 2015, pp. 2449–2457.

[47] H. W. Kuhn, “The hungarian method for the assignment problem,” *Naval Research Logistics Quarterly*, vol. 2, pp. 83–97, 1955.

[48] R. Labayrade, D. Aubert, and J.-P. Tarel, “Real time obstacle detection in stereovision on non flat road geometry through "v-disparity" representation,” in *IEEE Intelligent Vehicle Symposium (IV)*, 2002, pp. 646 – 651.

[49] M. Lauer, "A method to track vertical planes," unpublished work 2015.

[50] D. C. Lee, M. Hebert, and T. Kanade, "Geometric reasoning for single image structure recovery," in *IEEE Conference on Computer Vision and Pattern Recognition (CVPR)*, 2009.

[51] J.-K. Lee and K.-J. Yoon, "Real-time joint estimation of camera orientation and vanishing points," in *IEEE Conference on Computer Vision and Pattern Recognition (CVPR)*, 2015, pp. 1866–1874.

[52] Y. H. Lee, T.-S. Leung, and G. Medioni, "Real-time staircase detection from a wearable stereo system," in *International Conference on Pattern Recognition (ICPR)*, 2012, pp. 3770–3773.

[53] Y. H. Lee, C. Nam, K. Y. Lee, Y. S. Li, S. Y. Yeon, and N. L. Doh, "Vpass: Algorithmic compass using vanishing points in indoor environments," in *IEEE/RSJ International Conference on Intelligent Robots and Systems (IROS)*, 2009, pp. 936–941.

[54] P. Lenz, J. Ziegler, A. Geiger, and M. Roser, "Sparse scene flow segmentation for moving object detection in urban environments," in *IEEE Intelligent Vehicle Symposium (IV)*, 2011.

[55] T.-S. Leung and G. Medioni, "Visual navigation aid for the blind in dynamic environments," in *IEEE Conference on Computer Vision and Pattern Recognition (CVPR) Workshops*, 2014, pp. 579–586.

[56] A. Linarth, M. Brucker, and E. Angelopoulou, "Robust ground plane estimation based on particle filters," in *IEEE Intelligent Transportation Systems Conference (ITSC)*, 2009, pp. 1–7.

[57] P. Lombardi, M. Zanin, and S. Messelodi, "Unified stereovision for ground, road, and obstacle detection," in *IEEE Intelligent Vehicle Symposium (IV)*, 2005, pp. 783–788.

[58] X. Lu and R. Manduchi, "Detection and localization of curbs and stairways using stereo vision," in *IEEE International Conference on Robotics and Automation (ICRA)*, 2005, pp. 4648–4654.

[59] A. T. Martins, P. M. Q. Aguiar, and M. A. T. Figueiredo, "Orientation in manhattan: equiprojective classes and sequential estimation," *IEEE Transactions on Pattern Analysis and Machine Intelligence*, vol. 27, no. 5, pp. 822–827, 2005.

[60] S. Meers and K. Ward, "A substitute vision system for providing 3d perception and gps navigation via electro-tactile stimulation," in *International Conference on Sensing Technology*, 2005.

[61] P. B. Meijer, "The voice," http://www.seeingwithsound.com, accessed February 2017.

[62] F. M. Mirzaei and S. I. Roumeliotis, "Optimal estimation of vanishing points in a manhattan world," in *ICCV, Spain, November 6-13, 2011*, 2011, pp. 2454–2461.

[63] D. Mitzel and B. Leibe, "Taking mobile multi-object tracking to the next level: People, unknown objects, and carried items," in *European Conference on Computer Vision (ECCV)*, 2012, pp. 566–579.

[64] N. Molton, S. Se, M. Brady, D. Lee, and P. Probert, "Robotic sensing for the partially sighted," *Robotics and Autonomous Systems*, vol. 26, no. 2-3, pp. 185–201, 1999.

[65] "Project OIWOB," http://oiwob.de, accessed February 2017.

[66] A. Oliva and A. Torralba, "The role of context in object recognition," *Trends in Cognitive Sciences*, vol. 11, no. 12, pp. 520–527, 2007.

[67] S. Osswald, J. S. Gutmann, A. Hornung, and M. Bennewitz, "From 3d point clouds to climbing stairs: A comparison of plane segmentation approaches for humanoids," *IEEE-RAS International Conference on Humanoid Robots (Humanoids)*, pp. 93–98, 2011.

[68] A. Ošep, A. Hermans, F. Engelmann, D. Klostermann, M. Mathias, and B. Leibe, "Multi-scale object candidates for generic object tracking in street scenes," in *IEEE International Conference on Robotics and Automation (ICRA)*, 2016.

[69] A. Perez-Yus, G. Lopez-Nicolas, and J. J. Guerrero, "Detection and modelling of staircases using a wearable depth sensor," in *European Conference on Computer Vision (ECCV) Workshops*, 2014.

[70] S. Pillai and J. Leonard, "Monocular SLAM supported object recognition," in *Proceedings of Robotics: Science and Systems (RSS)*, 2015.

[71] V. Pradeep, G. Medioni, and J. Weiland, "Robot vision for the visually impaired," in *IEEE Conference on Computer Vision and Pattern Recognition (CVPR) Workshops*, 2010, pp. 15–22.

[72] V. Pradeep, G. Medioni, and J. Weiland, "Piecewise planar modeling for step detection using stereo vision," in *Workshop on Computer Vision Applications for the Visually Impaired*, 2008.

[73] B. Ranft and T. Strauß, "Modeling arbitrarily oriented slanted planes for efficient stereo vision based on block matching," in *IEEE Intelligent Transportation Systems Conference (ITSC)*, 2014.

[74] A. Rodríguez, J. J. Yebes, P. F. Alcantarilla, L. M. Bergasa, J. Almazán, and A. Cela, "Assisting the visually impaired: Obstacle detection and warning system by acoustic feedback," *Sensors*, vol. 12, no. 12, pp. 17 476–17 496, 2012.

[75] L. Ruotsalainen, J. Bancroft, and G. Lachapelle, "Mitigation of attitude and gyro errors through vision aiding," in *IEEE International Conference on Indoor Positioning and Indoor Navigation (IPIN)*, 2012, pp. 1–9.

[76] R. F. Salas-Moreno, R. A. Newcombe, H. Strasdat, P. H. J. Kelly, and A. J. Davison, "SLAM++: simultaneous localisation and mapping at the level of objects," in *IEEE Conference on Computer Vision and Pattern Recognition (CVPR)*, 2013, pp. 1352–1359.

[77] D. Scaramuzza and F. Fraundorfer, "Visual odometry [tutorial]," *Robotics Automation Magazine, IEEE*, vol. 18, no. 4, pp. 80–92, 2011.

[78] G. Schindler and F. Dellaert, "Atlanta world: an expectation maximization framework for simultaneous low-level edge grouping and camera calibration in complex man-made environments," in *IEEE Conference on Computer Vision and Pattern Recognition (CVPR)*, 2004, pp. 203–209.

[79] K. Schindler and H. Bischof, "On robust regression in photogrammetric point clouds." in *DAGM Symposium on Pattern Recognition*. Springer, 2003, pp. 172–178.

[80] S. Se and M. Brady, "Vision-based detection of kerbs and steps," in *British Machine Vision Conference (BMVC)*, 2000, pp. 410–419.

[81] S. Se and M. Brady, "Ground plane estimation, error analysis and applications." *Robotics and Autonomous Systems*, vol. 39, no. 2, pp. 59–71, 2002.

[82] S. Sengupta, E. Greveson, A. Shahrokni, and P. H. Torr, "Urban 3d semantic modelling using stereo vision," in *IEEE International Conference on Robotics and Automation (ICRA)*, 2013.

[83] S. Shoval, I. Ulrich, and J. Borenstein, "Navbelt and the guide-cane [obstacle-avoidance systems for the blind and visually impaired]," *Robotics and Automation Magazine, IEEE*, vol. 10, no. 1, pp. 9–20, 2003.

[84] G. Sibley, L. Matthies, and G. Sukhatme, *Robotics Research: Results of the 12th International Symposium ISRR*. Springer Berlin Heidelberg, 2007, ch. Bias Reduction and Filter Convergence for Long Range Stereo, pp. 285–294.

[85] D. Simon, *Optimal State Estimation: Kalman, H Infinity, and Nonlinear Approaches*. Wiley-Interscience, 2006.

[86] I. E. Sutherland, "Three-dimensional data input by tablet," *Proceedings of the IEEE*, vol. 62, no. 4, pp. 453–461, 1974.

[87] J.-P. Tardif, "Non-iterative approach for fast and accurate vanishing point detection," in *IEEE International Conference on Computer Vision (ICCV)*, 2009, pp. 1250–1257.

[88] R. Toldo and A. Fusiello, "Robust multiple structures estimation with j-linkage," in *European Conference on Computer Vision (ECCV)*, 2008, pp. 537–547.

[89] B. Triggs, P. F. McLauchlan, R. I. Hartley, and A. W. Fitzgibbon, *International Workshop on Vision Algorithms*. Springer Berlin Heidelberg, 2000, ch. Bundle Adjustment – A Modern Synthesis, pp. 298–372.

[90] V. Vineet, O. Miksik, M. Lidegaard, M. Nießner, S. Golodetz, V. A. Prisacariu, O. Kähler, D. W. Murray, S. Izadi, P. Pérez, and P. H. S. Torr, "Incremental dense semantic stereo fusion for large-scale semantic scene reconstruction," in *IEEE International Conference on Robotics and Automation (ICRA)*, 2015, pp. 75–82.

[91] C.-C. Wang, C. Thorpe, S. Thrun, M. Hebert, and H. Durrant-Whyte, "Simultaneous localization, mapping and moving object tracking," *International Journal of Robotics Research*, vol. 26, no. 9, pp. 889–916, 2007.

[92] S. Wang and H. Wang, "2d staircase detection using real adaboost," in *International Conference on Information, Communications and Signal Processing*, 2009, pp. 1–5.

[93] Y. Xu, S. Oh, and A. Hoogs, "A minimum error vanishing point detection approach for uncalibrated monocular images of man-made environments," in *IEEE Conference on Computer Vision and Pattern Recognition (CVPR)*, 2013, pp. 1376–1383.

[94] L. Zhang and R. Koch, "Vanishing points estimation and line classification in a manhattan world," in *Asian Conference on Computer Vision (ACCV)*, 2013, pp. 38–51.

[95] J. Zhou and B. Li, "Homography-based ground detection for a mobile robot platform using a single camera," in *IEEE International Conference on Robotics and Automation (ICRA)*, 2006, pp. 4100–4105.

[96] Óscar Martínez Mozos, A. Rottmann, R. Triebel, P. Jensfelt, and W. Burgard, "Supervised semantic labeling of places using information extracted from sensor data," *Robotics and Autonomous Systems*, vol. 55, pp. 391–402, 2007.

Publications by Author

[97] S. F. Amirhosseini, M. Romanovas, T. Schwarze, M. Schwaab, M. Traechtler, and Y. Manoli, "Stochastic cloning unscented kalman filtering for pedestrian localization applications," in *IEEE International Conference on Indoor Positioning and Indoor Navigation (IPIN)*, 2013, pp. 1–10.

[98] J. Gräter, T. Schwarze, and M. Lauer, "Robust scale estimation for monocular visual odometry using structure from motion and vanishing points," in *IEEE Intelligent Vehicle Symposium (IV)*, 2015.

[99] H. Harms, E. Rehder, T. Schwarze, and M. Lauer, "Detection of ascending stairs using stereo vision," in *IEEE/RSJ International Conference on Intelligent Robots and Systems (IROS)*, 2015, pp. 2496–2502.

[100] M. Romanovas, T. Schwarze, M. Schwaab, M. Traechtler, and Y. Manoli, "Stochastic cloning kalman filter for visual odometry and inertial/magnetic data fusion," in *IEEE International Conference on Information Fusion (FUSION)*, 2013, pp. 1434–1441.

[101] M. Schwaab, M. Romanovas, D. Plaia, T. Schwarze, and Y. Manoli, "Fusion of visual odometry and inertial sensors using dual quaternions and stochastic cloning," in *IEEE International Conference on Information Fusion (FUSION)*, 2016, pp. 573–580.

[102] T. Schwarze and M. Lauer, "Wall estimation from stereo vision in urban street canyons," in *International Conference on Informatics in Control, Automation and Robotics (ICINCO)*, 2013, pp. 83–90.

[103] T. Schwarze and M. Lauer, "Geometry estimation of urban street canyons using stereo vision from egocentric view," in *Informatics in Control, Automation and Robotics*, ser. Lecture Notes in Electrical Engineering. Springer International Publishing, 2015, vol. 325, pp. 279–292.

[104] T. Schwarze and M. Lauer, "Minimizing odometry drift by vanishing direction references," Presented at *International Conference on Indoor Positioning and Indoor Navigation (IPIN)*, 2015.

[105] T. Schwarze and M. Lauer, "Robust ground plane tracking in cluttered environments from egocentric stereo vision," in *IEEE International Conference on Robotics and Automation (ICRA)*, 2015, pp. 2442–2447.

[106] T. Schwarze, M. Lauer, M. Schwaab, M. Romanovas, S. Böhm, and T. Jürgensohn, "A camera-based mobility aid for visually impaired people," *KI - Künstliche Intelligenz*, pp. 1–8, 2015.

[107] T. Schwarze, M. Lauer, M. Schwaab, M. Romanovas, S. Böhm, and T. Jürgensohn, "An intuitive mobility aid for visually impaired people based on stereo vision," in *IEEE International Conference on Computer Vision (ICCV) Workshops*, 2015.

[108] T. Schwarze, M. Lauer, M. Schwaab, M. Romanovas, S. Böhm, and T. Jürgensohn, "Ein kamerabasierter Ansatz zur intuitiven Assistenz sehbehinderter Menschen," in *Forum Bildverarbeitung*, 2016, pp. 269–283.

[109] T. Schwarze, M. Lauer, M. Schwaab, M. Romanovas, S. Böhm, and T. Jürgensohn, "Ein intuitives kamerabasiertes System zur Assistenz sehbehinderter Menschen," vol. 84(7-8), pp. 535–545, 2017.

[110] T. Schwarze, M. Lauer, and C. Stiller, "OIWOB: Orientieren, Informieren, Warnen. Orientierungshilfe für Blinde : Abschlussbericht des Teilprojektes: 3D-Objekterkenung und semantische Analyse," Karlsruhe Institute of Technology, Tech. Rep., 2015.

[111] T. Schwarze and Z. Zhong, "Stair detection and tracking from egocentric stereo vision," in *IEEE International Conference on Image Processing (ICIP)*, 2015.

Supervised Theses

[112] A. Abderrahman, "Assistance systems for the visually impaired," Bachelor's thesis, Institut für Mess- und Regelungstechnik, KIT, 2015.

[113] R. Depoivre, "Object detection from egocentric viewpoints using u/v-disparity," Master's thesis, Institut für Mess- und Regelungstechnik, KIT, 2014.

[114] Y.-C. Huang, "Estimating camera orientation on running tracks in dynamic video sequences," Master's thesis, Institut für Mess- und Regelungstechnik, KIT, 2016.

[115] C. Kelegkouri, "Video segmentation based on multiple image cues," Master's thesis, Institut für Mess- und Regelungstechnik, KIT, 2015.

[116] C. Li, "Road marking recognition in panoramic photographs," Master's thesis, Institut für Mess- und Regelungstechnik, KIT, 2014.

[117] X. B. Sabaté, "Line tracking for stereoscopic video sequences," Master's thesis, Institut für Mess- und Regelungstechnik, KIT, 2015.

[118] X. Tan, "Planar patch classification based on multiple image cues," Master's thesis, Institut für Mess- und Regelungstechnik, KIT, 2013.

[119] Z. Zhong, "Stair detection and estimation from stereo vision," Master's thesis, Institut für Mess- und Regelungstechnik, KIT, 2014.

Schriftenreihe
Institut für Mess- und Regelungstechnik
Karlsruher Institut für Technologie
(1613-4214)

Band 001	Hans, Annegret **Entwicklung eines Inline-Viskosimeters auf Basis eines magnetisch-induktiven Durchflussmessers.** 2004 ISBN 3-937300-02-3
Band 002	Heizmann, Michael **Auswertung von forensischen Riefenspuren mittels automatischer Sichtprüfung.** 2004 ISBN 3-937300-05-8
Band 003	Herbst, Jürgen **Zerstörungsfreie Prüfung von Abwasserkanälen mit Klopfschall.** 2004 ISBN 3-937300-23-6
Band 004	Kammel, Sören **Deflektometrische Untersuchung spiegelnd reflektierender Freiformflächen.** 2005 ISBN 3-937300-28-7
Band 005	Geistler, Alexander **Bordautonome Ortung von Schienenfahrzeugen mit Wirbelstrom-Sensoren.** 2007 ISBN 978-3-86644-123-1
Band 006	Horn, Jan **Zweidimensionale Geschwindigkeitsmessung texturierter Oberflächen mit flächenhaften bildgebenden Sensoren.** 2007 ISBN 978-3-86644-076-0

Band 007	Hoffmann, Christian **Fahrzeugdetektion durch Fusion monoskopischer Videomerkmale.** 2007 ISBN 978-3-86644-139-2
Band 008	Dang, Thao **Kontinuierliche Selbstkalibrierung von Stereokameras.** 2007 ISBN 978-3-86644-164-4
Band 009	Kapp, Andreas **Ein Beitrag zur Verbesserung und Erweiterung der Lidar-Signalverarbeitung für Fahrzeuge.** 2007 ISBN 978-3-86644-174-3
Band 010	Horbach, Jan **Verfahren zur optischen 3D-Vermessung spiegelnder Oberflächen.** 2008 ISBN 978-3-86644-202-3
Band 011	Böhringer, Frank **Gleisselektive Ortung von Schienenfahrzeugen mit bordautonomer Sensorik.** 2008 ISBN 978-3-86644-196-5
Band 012	Xin, Binjian **Auswertung und Charakterisierung dreidimensionaler Messdaten technischer Oberflächen mit Riefentexturen.** 2009 ISBN 978-3-86644-326-6
Band 013	Cech, Markus **Fahrspurschätzung aus monokularen Bildfolgen für innerstädtische Fahrerassistenzanwendungen.** 2009 ISBN 978-3-86644-351-8
Band 014	Speck, Christoph **Automatisierte Auswertung forensischer Spuren auf Patronenhülsen.** 2009 ISBN 978-3-86644-365-5

Die Bände sind unter www.ksp.kit.edu als PDF frei verfügbar oder als Druckausgabe bestellbar.

Band 015	Bachmann, Alexander **Dichte Objektsegmentierung in Stereobildfolgen.** 2010 ISBN 978-3-86644-541-3
Band 016	Duchow, Christian **Videobasierte Wahrnehmung markierter Kreuzungen mit lokalem Markierungstest und Bayes'scher Modellierung.** 2011 ISBN 978-3-86644-630-4
Band 017	Pink, Oliver **Bildbasierte Selbstlokalisierung von Straßenfahrzeugen.** 2011 ISBN 978-3-86644-708-0
Band 018	Hensel, Stefan **Wirbelstromsensorbasierte Lokalisierung von Schienenfahrzeugen in topologischen Karten.** 2011 ISBN 978-3-86644-749-3
Band 019	Carsten Hasberg **Simultane Lokalisierung und Kartierung spurgeführter Systeme.** 2012 ISBN 978-3-86644-831-5
Band 020	Pitzer, Benjamin **Automatic Reconstruction of Textured 3D Models.** 2012 ISBN 978-3-86644-805-6
Band 021	Roser, Martin **Modellbasierte und positionsgenaue Erkennung von Regentropfen in Bildfolgen zur Verbesserung von videobasierten Fahrerassistenzfunktionen.** 2012 ISBN 978-3-86644-926-8

Die Bände sind unter www.ksp.kit.edu als PDF frei verfügbar oder als Druckausgabe bestellbar.

Band 022 Loose, Heidi
Dreidimensionale Straßenmodelle für Fahrerassistenzsysteme auf Landstraßen. 2013
ISBN 978-3-86644-942-8

Band 023 Rapp, Holger
Reconstruction of Specular Reflective Surfaces using Auto-Calibrating Deflectometry. 2013
ISBN 978-3-86644-966-4

Band 024 Moosmann, Frank
Interlacing Self-Localization, Moving Object Tracking and Mapping for 3D Range Sensors. 2013
ISBN 978-3-86644-977-0

Band 025 Geiger, Andreas
Probabilistic Models for 3D Urban Scene Understanding from Movable Platforms. 2013
ISBN 978-3-7315-0081-0

Band 026 Hörter, Marko
Entwicklung und vergleichende Bewertung einer bildbasierten Markierungslichtsteuerung für Kraftfahrzeuge. 2013
ISBN 978-3-7315-0091-9

Band 027 Kitt, Bernd
Effiziente Schätzung dichter Bewegungsvektorfelder unter Berücksichtigung der Epipolargeometrie zwischen unterschiedlichen Ansichten einer Szene. 2013
ISBN 978-3-7315-0105-3

Band 028 Lategahn, Henning
Mapping and Localization in Urban Environments Using Cameras. 2013
ISBN 978-3-7315-0135-0

Band 029 Tischler, Karin
Informationsfusion für die kooperative Umfeldwahrnehmung vernetzter Fahrzeuge. 2014
ISBN 978-3-7315-0166-4

Band 030 Schmidt, Christian
Fahrstrategien zur Unfallvermeidung im Straßenverkehr für Einzel- und Mehrobjektszenarien. 2014
ISBN 978-3-7315-0198-5

Band 031 Firl, Jonas
Probabilistic Maneuver Recognition in Traffic Scenarios. 2014
ISBN 978-3-7315-0287-6

Band 032 Schönbein, Miriam
Omnidirectional Stereo Vision for Autonomous Vehicles. 2015
ISBN 978-3-7315-0357-6

Band 033 Nicht erschienen

Band 034 Liebner, Martin
Fahrerabsichtserkennung und Risikobewertung für warnende Fahrerassistenzsysteme. 2016
ISBN 978-3-7315-0508-2

Band 035 Ziegler, Julius
Optimale Trajektorienplanung für Automobile. 2017
ISBN 978-3-7315-0553-2

Band 036 Harms, Hannes
Genauigkeitsuntersuchung von binokularen Normalenvektoren für die Umfeldwahrnehmung. 2017
ISBN 978-3-7315-0628-7

Band 037 Ruhhammer, Christian
Inferenz von Kreuzungsinformationen aus Flottendaten. 2017
ISBN 978-3-7315-0721-5

Band 038 Stein, Denis
Mobile laser scanning based determination of railway network topology and branching direction on turnouts. 2018
ISBN 978-3-7315-0743-7

Band 039 Yi, Boliang
Integrated Planning and Control for Collision Avoidance Systems. 2018
ISBN 978-3-7315-0785-7

Band 040 Schwarze, Tobias
Compact Environment Modelling from Unconstrained Camera Platforms. 2018
ISBN 978-3-7315-0801-4